QUANTUM MECHANICS
A Complete Introduction

Alexandre Zagoskin is a Reader in Quantum Physics at Loughborough University and a Fellow of The Institute of Physics.

He received his Master's degree in physics from Kharkov University (Soviet Union), and his PhD in physics from the Institute for Low Temperature Physics and Engineering (FTINT), also in Kharkov. He has worked at FTINT, Chalmers University of Technology (Sweden), the University of British Columbia (Canada) and the RIKEN Institute (Japan). In 1999, when in Canada, he co-founded the company D-Wave Systems with the aim of developing commercial quantum computers, and was its Chief Scientist and VP of Research until 2005.

He has published two books: Quantum Theory of Many-Body Systems *(Springer, 1998, 2014) and* Quantum Engineering *(Cambridge University Press, 2011).*

He and his colleagues are now working on the development of quantum engineering – the theory, design, fabrication and characterization of quantum coherent artificial structures.

QUANTUM MECHANICS
A Complete Introduction

Alexandre Zagoskin

Teach® Yourself

Contents

How to use this book

This Complete Introduction from Teach Yourself® includes a number of special boxed features, which have been developed to help you understand the subject more quickly and remember it more effectively. Throughout the book, you will find these indicated by the following icons.

 The book includes concise **quotes** from other key sources. These will be useful for helping you understand different viewpoints on the subject, and they are fully referenced so that you can include them in essays if you are unable to get your hands on the source.

 The **case study** is a more in-depth introduction to a particular example. There is at least one in most chapters, and they should provide good material for essays and class discussions.

 The **key ideas** are highlighted throughout the book. If you only have half an hour to go before your exam, scanning through these would be a very good way of spending your time.

 The **spotlight** boxes give you some light-hearted additional information that will liven up your learning.

 The **fact-check** questions at the end of each chapter are designed to help you ensure you have taken in the most important concepts from the chapter. If you find you are consistently getting several answers wrong, it may be worth trying to read more slowly, or taking notes as you go.

 The **dig deeper** boxes give you ways to explore topics in greater depth than is possible in this introductory-level book.

Acknowledgements

I am grateful to my friends, colleagues and students, discussions with whom helped me better understand many topics covered in this book; and to D. Palchun, R. Plehhov and E. Zagoskina, who agreed to be test subjects for some parts of the manuscript and lived to tell me about the experience.

Introduction

A fair warning

> The first thing a student of magic learns is that there are books about *magic* and books *of* magic. And the second thing he learns is that a perfectly respectable example of the former may be had for two or three guineas at a good bookseller, and that the value of the latter is above rubies.
>
> Susanna Clarke, *Jonathan Strange & Mr Norrell*

This book will help you teach yourself *about* quantum mechanics, which is not quite the same as teaching yourself quantum mechanics. Here is the difference.

To learn quantum mechanics means being able to *do* something with quantum mechanics as a matter of course (solving equations, mostly. If you are an experimentalist, this also entails conducting experiments, the planning and analysis of which require solving equations), and to have been doing this long enough to acquire a certain intuition about these equations and experiments – and therefore, about quantum mechanics. At the very least, one would have to go through a good undergraduate textbook on quantum mechanics and solve all the problems (which means also going through a couple of textbooks on mathematics and solving the problems from them too), and discuss all the questions you may have with somebody who already has the above-mentioned intuition. In other words, to take a regular introductory course of quantum mechanics at a university, and this is not always an option. And anyway, you may not be planning to become a professional physicist (which is a pity).

You can instead learn *about* quantum mechanics. This is a book written exactly for this purpose – to help you teach yourself *about* quantum mechanics. This means that in the end you will not become a practising magician, that is, physicist – but hopefully you will have a better understanding of what physicists do and think when they do research in quantum mechanics, and what this research tells us about the world we live in. Besides, much of our technology uses quantum mechanical effects routinely (like lasers and semiconductor microchips, which were produced by the 'first quantum revolution' back in the 20th century), and if the current trends hold, we will be soon enough playing with the technological results of the 'second quantum revolution', using even more subtle and bizarre quantum effects. It is therefore wise to be prepared. As a bonus, you will learn about quite a lot of things that are *not* taught in introductory quantum mechanics courses to physics undergraduates.

This book contains quite a number of equations (though much less than a book *of* quantum mechanics would), so brushing up your school maths would be a good idea. I have tried to introduce all the new mathematical things, and some things you may already know. Still, this is just a book *about* quantum mechanics. So why equations?

Quantum mechanics describes our world with an unbelievable precision. But the scale of phenomena, where this description primarily applies, is so far removed from our everyday experience that our human intuition and language, however rich, are totally inadequate for the task of expressing quantum mechanics directly.

In such straits it was common to use poetry with its allusions, allegories and similes. This is indeed the only way one can write about quantum mechanics using only words – even if the poetry lacks rhymes and rhythm. This is a nice and often inspiring way of doing things, but poetry is too imprecise, too emotional and too individual.

It is therefore desirable to use the language in which quantum mechanics *can* be expressed: mathematics, a language more remote from our common speech than any Elvish dialect. Fortunately, this is also the language in which *all* physics is most naturally expressed, and it applies to such areas – like mechanics – where we do have an inborn and daily trained intuition. I did therefore try to relate the one to the other – you will be the judge of how well this approach succeeded.[1]

---

1 The more 'mathematical' sections, which can be skipped at a first reading, are marked with an asterisk.

1

Familiar physics in strange spaces

To an unaccustomed ear, the very language of quantum mechanics may seem intimidatingly impenetrable. It is full of commutators, operators, state vectors, Hilbert spaces and other mathematical horrors. At any rate quantum physics seems totally different from the clarity and simplicity of Newtonian mechanics, where footballs, cars, satellites, stars, planets, barstools, bees, birds and butterflies move along well-defined trajectories, accelerate or decelerate when acted upon by forces, and always have a definite position and velocity. All this can be described by the simple mathematical laws discovered by Sir Isaac Newton and taught at school – or just intuitively comprehended based on our everyday experience. Quantum mechanics, on the contrary, is best left to somebody else.

Spotlight: Rene Descartes (1596–1650)

As the tradition has it, Descartes discovered the Cartesian coordinates when serving as a military officer. He had a lazy day indoors, observed a fly wandering on the ceiling and realized that its position can be determined by its distance from the walls.

Why 'Cartesian'? Descartes wrote some of his works in Latin, and his Latinized name is Cartesius.

Actually, this impression is false. We are not saying that quantum mechanics does not differ from classical mechanics – of course it does, and it indeed presents a counterintuitive, puzzling, very beautiful and very precise view of the known Universe. But to see these differences in the proper light, it is necessary first to cast a better look at the structure of classical mechanics and realize – to our surprise – that in many important respects it is very similar to quantum mechanics.

Key idea: Coordinate transformation

Switching between different coordinate systems is called *coordinate transformation*. Here is a simple example. Suppose a point A has coordinates (x,y,z). Let us now introduce a new coordinate system $(O'x'y'z')$ with the origin in the point O' with coordinates (X,Y,Z).

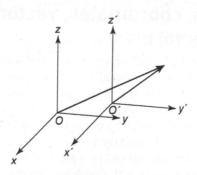

As you can see from the diagram, the coordinates (x',y',z') of the same point A in the new system are given by

$$x' = x - X, y' = y - Y, z' = z - Z.$$

Any coordinate transformation can be written as:

$$(x',y',z') = \Lambda (x,y,z).$$

Here Λ is shorthand for an explicitly known set of operations one must perform with the set of coordinates in the original system, in order to obtain the coordinates of the same point in the new one. Mathematically, this means that Λ is an *operator* – the first scary word is demystified.

Of course, to see this one has to venture beyond Newtonian dynamics, but this exercise does not require any extraordinary effort. We will however need some mathematics. The reason being that mathematics is a very concise and precise language, specially developed over the centuries to describe complex natural phenomena and to make precise and verifiable statements. Physics in this respect is not unlike Shakespeare's works: in order to enjoy the play and understand what it is about, you should know the language, but you do not need to know the meaning of every single word or be good at writing sonnets.

Much of what will be required for the following you may have already studied at school (but perhaps looked at from a different angle) – look at the boxes for reminders, explanations and other useful information.

Now let us start with some basics.

Particles, coordinates, vectors and trajectories

Consider a *material point* or a *particle* – the simplest object of Newtonian dynamics. Visualize this as an infinitesimally small, massive, structureless object. For example, a good model for this would be a planet orbiting the Sun but not too close to it, or the bob of a pendulum. Since the times of Descartes, we know that the position of a point in space can be completely determined by a set of three numbers – its *Cartesian coordinates* (x,y,z), which give its distances from the origin O along the three mutually orthogonal (that is, at right angles to each other) coordinate axes Ox, Oy, Oz (Figure 1.1). These axes and their common origin O (*the reference point*) form a Cartesian *coordinate system*. Just look at a corner of your room, and you will see, as did Descartes four centuries ago, the origin and the three coordinate axes meeting there.[2]

Spotlight: At a tangent

According to classical mechanics, 'to fly at a tangent' actually means 'to keep moving in the *same* direction as before', and not 'to start on something completely different'!

Key idea: Functions

A quantity f, which depends on some variable x, is called a *function* of x and is usually denoted by $f(x)$. Thus, the radius vector is a function of x, y and z.

---

2   If you happen to live in a round, spherical or even crooked house, there exist appropriate coordinate systems too – but they would lead us too far away from our purpose.

Spotlight: Looking for treasure

If you wish, you can treat a radius vector as a set of instructions on a treasure map. For example, from the old oak tree, go *x* feet east, then *y* feet north, then dig a hole (-*z*) feet deep (or climb *z* feet up a cliff).

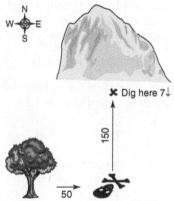

There can be any number of such coordinate systems, with different reference points and different directions of the axes, but once the coordinates (*x*,*y*,*z*) of a point are known in one coordinate system, they are readily found in any other system.

We can also determine the position of our point by a *radius vector r* – a straight arrow connecting it with the reference point *O* (Figure 1.1). Of course, both descriptions are equivalent, and we can write simply

$$\mathbf{r} = (x, y, z),$$

that is, a vector is a set of three Cartesian coordinates (which are called its *Cartesian components* and must be properly transformed if we decide to switch to another coordinate system). We can easily add or subtract two vectors by adding or subtracting their corresponding components. For example, if our material point moved the distance Δx along the axis Ox, Δy along Oy, and Δz along Oz, its new position will be given by a radius vector (Figure 1.2):

$$r' = (x + \Delta x, y + \Delta y, z + \Delta z) \equiv r + \Delta r,$$

where the vector $\Delta r = (\Delta x, \Delta y, \Delta z)$ is the displacement.

(The sign of equivalence '$\equiv$' means 'the same as', and by Δa one customarily denotes a change in the variable a.)

If we now add to our coordinate system a clock, which measures time t, it becomes a *reference frame*. The position of a point will now depend on time,

$$r(t) \equiv (x(t), y(t), z(t)),$$

and as the time passes, the end of the radius vector traces a line in space – the *trajectory* of the point (Figure 1.1).

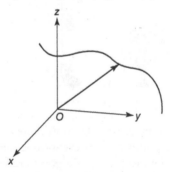

Figure 1.1 Cartesian coordinates, radius vector and a trajectory.

Velocity, acceleration, derivatives and differential equations

Key idea: Calculus

The calculus, including the concept of *differentiation* (finding derivatives), was independently developed by Newton and Leibniz. The nomenclature and notation proposed by Leibniz (in particular, the term 'derivative' and the symbol $\frac{dx}{dt}$ for it) turned out to be more convenient, although sometimes Newtonian notation is also used:

$$\dot{x} \equiv \frac{dx}{dt}$$

Key idea: Derivative

The derivative $\frac{df}{dx}$ is the mathematically rigorous expression for the rate of change of a function $f(x)$ as the variable x changes. For example, the incline of a slope is given by the derivative of height with respect to the horizontal displacement:

$$\tan \alpha = \frac{dh(x)}{dx}$$

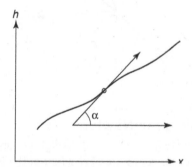

The Leibniz's notation for the derivative reflects the fact that it is a limit of the ratio of a vanishingly small change Δf, to a vanishingly small change Δx.

Now we can define the *velocity*. Intuitively, the velocity tells us how fast something changes its position in space. Let the point move for a short time interval Δt from the position $r(t)$ to $r(t + \Delta t)$ (Figure 1.2):

$$r(t + \Delta t) = r(t) + \Delta r.$$

Its average velocity is then $v_{av} = \frac{\Delta r}{\Delta t}$. Now, let us make the time interval Δt smaller and smaller (that is, take the limit at $t \rightarrow 0$). The average velocity will change. But, if the trajectory is smooth enough – and we will not discuss here those crazy (but sometimes important) special cases, of which mathematicians are so fond – then the average velocity will get closer and closer to a fixed vector $v(t)$. This vector is called the velocity (or the instantaneous velocity) of our point at the moment t:

$$v(t) = \lim_{\Delta t \to 0} \frac{\Delta r}{\Delta t} \equiv \frac{dr}{dt}.$$

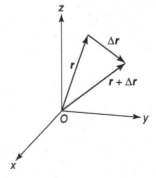

Figure 1.2 Displacement

Spotlight: Sir Isaac Newton

Sir Isaac Newton (1642–1727) was not knighted for his great discoveries, or for his role as the President of the Royal Society, or even for his exceptional service to the Crown as the Warden and then Master of the Royal Mint (he organized the reissuing of all English coinage within 2 years and under budget, crushed a ring of counterfeiters resulting in their ringleader being hanged, and introduced new equipment and new safety measures against counterfeiting: a ribbed rim with the *Decus et tutamen* (Latin for 'Ornament and safeguard') inscription, which has marked the British pound coin ever since). Newton's promotion was due to an attempt by his political party to get him re-elected to Parliament in order to keep the majority and retain power. The attempt failed in all respects – but Newton received his well-deserved knighthood.

In mathematical terms, the velocity is the *derivative* (or the *first* derivative, to be more specific) of the radius vector with respect to time:

$$v \equiv \left(v_x, v_y, v_z\right) = \frac{dr}{dt} \equiv \left(\frac{dx}{dt}, \frac{dy}{dt}, \frac{dx}{dt}\right).$$

The velocity $v(t)$ is *always* at a tangent to the trajectory $r(t)$, as you can see from Figure 1.3.

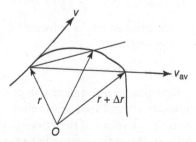

Figure 1.3 Average and instantaneous velocity

The rate of change of the velocity – the *acceleration* – can be found in the same way. It is the first derivative of velocity with respect to time. Since velocity is already the derivative of radius vector, the acceleration is the derivative of a derivative, that is, the *second derivative* of the radius vector with respect to time:

$$a(t) = \frac{dv}{dt} = \frac{d}{dt}\left(\frac{dr}{dt}\right) = \frac{d^2r}{dt^2}$$

(Note that it is d^2r/dt^2 and not "d^2r/d^2t"!).

The same equations can be written in components:

$$\left(a_x, a_y, a_z,\right) = \left(\frac{dv_x}{dt}, \frac{dv_y}{dt}, \frac{dv_z}{dt}\right) = \left(\frac{d^2x}{dt^2}, \frac{d^2y}{dt^2}, \frac{d^2z}{dt^2}\right).$$

We could spend the rest of our lives calculating third, fourth and so on derivatives, but fortunately we do not have to. Thanks to Newton, we only need to deal with position, velocity and acceleration of a material point.

Here is the reason why. Newton's second law of motion relates the acceleration a, of a material point with mass m with a force F acting on it:

$$ma = F.$$

This is how this law is usually written in school textbooks. Newton himself expressed it in a somewhat different form (in modern-day notation):

$$\frac{dp}{dt} = F.$$

Here we introduce the product of the velocity by the mass of the material point – the *momentum*:

$$p = mv,$$

which is another important vector quantity in classical (and quantum) mechanics, as we will see soon enough. Since the acceleration is the time derivative of the velocity, and the velocity is the time derivative of the radius vector, both these ways of writing Newton's second law are equivalent. We can even write it as $m\frac{d^2r}{dt^2} = F\left(r(t), \frac{dr}{dt}\right)$. Here we have stressed that the position r depends on time. It is this dependence (that is, where our material point – a football, a planet, a bullet – will be at a given moment) that we want to find from Newton's second law.

The force F depends on position and – sometimes – on the velocity of the material point. In other words, it is a known function of both r and dr/dt. For example, if we place the origin O of the frame of reference in the centre of the Sun, then the force of gravity attracting a planet to the Sun is $F = -\frac{GM_\odot m}{r^3}r$. Here G is the universal gravity constant, $M_\odot$ is the mass of the Sun, and m is the mass of the planet. Another example is the force of resistance, acting on a sphere of radius R, which slowly moves through a viscous liquid: $F = -6\pi\eta Rv$, where η is the viscosity of the liquid and v is the velocity of the sphere.

The latest version of the second law has the advantage of making clear that it is what mathematicians call a *second order differential equation* for a function $r(t)$ of the independent variable t (because it includes the second derivative of $r(t)$, but no derivatives of the higher order). Solving such equations is not necessarily easy, but is – in principle – possible, if only we know the *initial conditions* – the position $r(0)$ and the velocity $v(0) \equiv dr/dt$ ($t = 0$) at the initial moment. Then one can – in principle – predict the position $r(t)$ (and therefore the velocity) at *any future moment of time.*[3] Since we do not require any higher-order derivatives of the radius vector in order to make this prediction, we do not bother calculating them. Positions, velocities and accelerations are everything we need.

Phase space

Now here surfaces a little problem with our picture of a trajectory in real space, Figure 1.1. Unless we somehow put time stamps along the trajectory, we cannot tell how fast – or even in what direction! – a particle moved at any point, and will be, of course, unable to tell how it will behave after 'flying off at a tangent'. Such time stamps are messy and, frankly, no big help (Figure 1.4). Using them would be like a forensics unit trying to figure out what happened during a road accident based on the skid tracks on the road alone.

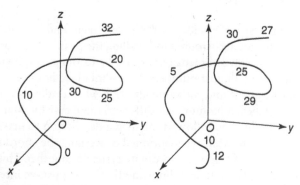

Figure 1.4 'Time stamps' along a trajectory

However, an improvement was suggested long ago. What if we doubled the number of coordinate axes, adding an extra axis for each velocity (or, rather, momentum) component? Then we would be able to plot simultaneously the position and the momentum of our material point. It will be enough to look at the picture to see not only where it is, but where to and how

---

3　This is an example of so-called *Laplacian determinism*: in classical physics, the future to the end of times is *completely determined* by the present.

fast it moves. Thus the motion of our material point will be described not in a three-dimensional real space with coordinates (x,y,z), but in a six-dimensional space with coordinates (x,p_x,y,p_y,z,p_z). Such space is called the *phase space* (Figure 1.5).

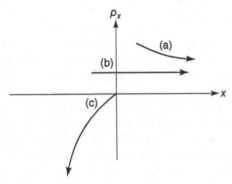

Figure 1.5 Phase space

Why call it a 'space'? Well, it is both convenient and logical. We have already got accustomed to labelling the position of a point in our convenient, 3-dimensional space by its three coordinates, which are, after all, just numbers. Why should we not use the same language and the same intuition when dealing with something labelled by more (or different) numbers? Of course, certain rules must be followed for such a description to make sense – similar to what we mentioned concerning the coordinate transformations from one coordinate system to another. This is somewhat similar to first calling 'numbers' only positive integers, which can be used for counting; then introducing fractions; then irrational numbers, like the famous π; then the negative numbers; then the imaginary ones, and so on. Of course, 'new numbers' have some new and unusual properties, but they are still numbers, and what works for 1, 2, 3 works for 22/7, 3.14159265358..., −17 or $\sqrt{-1}$. Moreover, the reason behind introducing each new kind of number is to make our life simpler, to replace many particular cases by a single general statement.

Now let us return to our six-dimensional phase space and ask: what kind of simplification is this? Can one even *imagine* a six-dimensional space? The answer is yes. We can start practising with a simple example of a phase space – so simple, that it can be drawn on a piece of paper.

Let us consider a particle, which can only move along one axis – like a bob of a spring pendulum, a fireman sliding down a pole or a train carriage on a straight section of the tracks. Then the only coordinate that can change is, say, x, and the rest do not matter (it is customary to denote the vertical coordinate by z, but this is more like a guideline). Now we can plot the corresponding momentum, px, along an axis orthogonal to Ox, and obtain a quite manageable, two-dimensional phase space (Figure 1.5). The negative values of px mean that the particle moves in the negative direction of the Ox axis. The trajectory of a particle in this space is easy to read. For example, particle (a) started moving to the right along Ox with a finite speed and slows down, (b) moves right with constant velocity, and (c) was initially at rest and then started accelerating to the left. On the other hand, you can easily see that – unlike the trajectories in real space – inverting the direction along these trajectories is impossible: a particle cannot move left, if its velocity points right, etc. This indicates a special structure of the phase space, which is indeed very important and useful, but we do not have to go into this.

Key idea: Sine and cosine

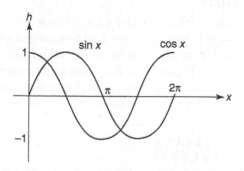

Sine and cosine functions describe oscillations and have the following properties:

$$\sin^2 \omega t + \cos^2 \omega t = 1;$$

$$\frac{d}{dt}\sin \omega t = \omega \cos \omega t;$$

$$\frac{d}{dt}\cos \omega t = -\omega \sin \omega t.$$

Therefore you can check that the differential equation

$$\frac{d^2 f}{dt^2} = -\omega^2 f(t)$$

has a general solution $f(t) = A \cos\omega t + B \sin \omega t$.

Spotlight: Robert Hooke (1635–1703)

A great English scientist, architect and inventor, who was probably the only man whom Newton considered as a serious rival in science. Hooke's law – that applies to any elastic forces produced by small deformations – is the best known and, arguably, the least significant of his contributions to science.

Harmonic oscillator, phase trajectories, energy conservation and action

Now let us move on and consider a special case of a system with a two-dimensional phase space – a one-dimensional *harmonic oscillator*, which is an idealized model of a simple pendulum, a spring pendulum (Figure 1.6), and many other important physical systems (such as atoms, molecules, guitar strings, clocks, bridges and anything whatsoever that can oscillate).

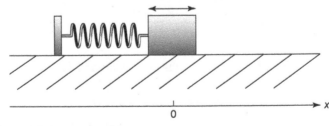

Figure 1.6 A spring pendulum

Here the force (e.g. the reaction of the spring) is proportional to the deviation from it (*Hooke's law*): $F(x) = -kx$.

The number k is the rigidity of the spring, x is the coordinate of the bob, and we chose the origin at the point of equilibrium. The minus sign simply indicates that the elastic force tends to return the bob back to the equilibrium position.

Newton's second law for the bob is then written as:

$$m\frac{d^2x}{dt^2} = -kx,$$

$$\text{or } \frac{d^2x}{dt^2} = -\omega_0^2 x.$$

In the latter formula we have introduced the quantity $\omega_0 = \sqrt{k/m}$, which is determined by the spring's rigidity and the particle mass. What is its meaning? Well, the above equation had been solved centuries ago, and the answers are known and given by any combination of sines and cosines:

$$x(t) = A\cos\omega_0 t + B\sin\omega_0 t.$$

We see that this is indeed an oscillating function, and ω_0 determines the frequency of oscillations. It is called the *oscillator's own frequency*. Its dependence on k and m makes sense; the heavier the bob, and the weaker the spring, the slower are the oscillations.

Using the properties of trigonometric functions, we can now immediately write:

$$v(t) \equiv \frac{dx(t)}{dt} = -A\omega_0 \sin\omega_0 t + \omega_0 B\cos\omega_0 t.$$

Here A and B are some constants, which are determined by the initial position, x_0, and velocity, v_0:

$$x_0 = x(0) = A; v_0 = v(0) = \omega_0 B.$$

Now we can write the equations, which determine the *phase trajectories* of our oscillator:

$$x(t) = x_0 \cos \omega_0 t + \frac{p_{x0}}{m\omega_0} \sin \omega_0 t;$$

$$p_x(t) = -m\omega_0 x_0 \sin \omega_0 t + p_{x0} \cos \omega_0 t,$$

and plot them for different values of the initial position and momentum, x_0 and $p_{x0} \equiv m v_0$ (Figure 1.7). It is enough to trace the trajectory for time $0 \le t \le T_0 \equiv \frac{2\pi}{\omega_0}$. After that the motion will repeat itself, as it should: this is an oscillator, so it oscillates with the period T_0. (The frequency ω_0 is the so-called *cyclic* frequency and is measured in inverse seconds. More familiar to non-physicists is the *linear* frequency $v_0 \equiv 1/T_0 = \omega_0/2\pi$, which tells us how many oscillations per second the system does, and is measured in Hertz (Hz): 1Hz is one oscillation per second).

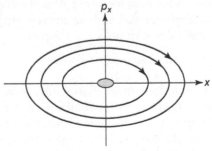

Figure 1.7 Phase trajectories of a harmonic oscillator

 Spotlight: Analytic geometry

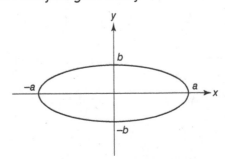

One of the great achievements of Descartes was the development of analytic geometry. In other words, he discovered how to translate from the language of curves and shapes (geometry) to the language of equations (algebra) and back again. In particular, it turned out that ellipse, one of special curves known already to ancient Greeks, is described by an equation

$$\frac{x^2}{a^2} + \frac{y^2}{b^2} = 1.$$

Here, if $a > b$, a is the semi-major axis and b is the semi-minor axis. The area of an ellipse is

$$S = \pi ab.$$

A circle of radius R is a special case of an ellipse, with $a = b = R$.

Key idea: Mechanical energy

Mechanical energy characterizes the ability of a given system to perform work. Kinetic energy is determined by the velocities, while the potential energy depends on the positions of the system's parts.

First, we see that all the trajectories are closed curves; second, that the motion along them is periodic with the same period (this is why the oscillator is called 'harmonic': if it were a string, it would produce the *same* tune no matter *how or where it was* plucked); and, third, that these trajectories fill the phase space – that is, there is one and only one (as the mathematicians like to say) trajectory passing through any given point. In other words, for any combination of initial conditions (x_0, p_{x0}) there is a solution – which is clear from its explicit expression – and this solution is completely determined by these initial conditions.

What is the shape of the curves in Figure 1.7? Recalling the basic properties of trigonometric functions, we see that

$$\frac{x^2}{x_0^2 \left[1 + \left(\dfrac{p_{x0}}{m\omega_0 x_0} \right)^2 \right]} + \frac{p_x^2}{(m\omega_0 x_0)^2 \left[1 + \left(\dfrac{p_{x0}}{m\omega_0 x_0} \right)^2 \right]} = 1.$$

Due to Descartes and his analytic geometry, we know that this equation describes an ellipse with the semi-axes

$$x_m = x_0\sqrt{1+\left(\frac{px_0}{m\omega_0 x_0}\right)^2} \text{ and } p_m = m\omega_0 x_0\sqrt{1+\left(\frac{px_0}{m\omega_0 x_0}\right)^2}.$$

Spotlight: Integration

Newton and Leibniz discovered integration independently, but we are all using Leibniz's notations, as with the derivative.

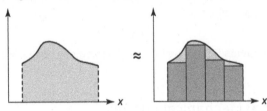

Let us consider a function $f(x)$ and find the area enclosed by its graph and the axis Ox, between the points $x = a$ and $x = b$. To do so, we can slice the interval $[a,b]$ in a large number N of small pieces of length $\Delta x = \frac{b-a}{N}$ and approximate the area by the sum of the areas of narrow rectangles of width Δx and height $f(x_i)$:

$$S \approx \sum_{i=1}^{N} f\left(x_i\right)\Delta x$$

When the width of rectangles becomes infinitely small, and their number infinitely large, the sum becomes *the integral of f(x) over x in the limits a, b*:

$$\int_a^b f(x)dx = \lim_{\substack{\Delta x \to 0 \\ N \to \infty}} \sum f(x_i)\Delta x$$

The sign of integral is simply a stylized 'S' (from 'summa', the sum), and dx is a symbol for Δx, when $\Delta x \to 0$.

Let us look again at the above equation. It tells us that some combination of coordinate and momentum is the same for any point on the trajectory, that is, is *conserved* during the motion. Recalling that the potential energy of a spring is

$U(x) = kx^2/2 = m\omega_0^2 x^2/2$, and the kinetic energy of the bob is $K = mv^2/2 = p_x^2/2m$, we can easily check that the equation for the elliptic phase trajectory is nothing else but the *mechanical energy conservation law*,

$$H(x, p_x) = U(x) + K(p_x) = \frac{m\omega_0^2 x^2}{2} + \frac{p_x^2}{2m} = \text{constant}.$$

We have arrived at it through a rather unusual route – via the phase space. Before explaining why I denote energy with the capital H, let us introduce another important physical quantity. Each of the ellipses in Figure 1.7 can be uniquely characterized by its own area. In our case

$$S = \pi x_m p_m = \pi m\omega_0 x_0^2 \left[1 + \left(\frac{p_{xo}}{m\omega_0 x_0} \right)^2 \right] = \frac{\pi}{\omega_0} \left[m\omega_0^2 x_0^2 + \frac{p_{x0}^2}{m\omega_0} \right]$$

$$= \frac{2\pi}{\omega_0} H(x_0, p_{x0}) \equiv T_0 H(x_0, p_{x0}).$$

The quantity S is called *action*, and we will have many opportunities to see the important role it plays in both classical and quantum mechanics. Suffice to mention here that the celebrated *Planck constant h* is nothing else but the *quantum of action* – the smallest change of action one can ever get. But let us not run ahead.

For a harmonic oscillator the action is simply the product of the oscillator energy and its period of oscillation. In a more general case – for example, if our spring does not quite follow Hooke's law, and its elastic force is not simply proportional to the displacement – the phase trajectories are not elliptic and have different periods of oscillations. In any case, the action – that is, the area enclosed by a phase trajectory – can always be found from the formula:

$$S = \oint p_x(x) dx.$$

Here x is the coordinate, $p_x = mv_x$ is the corresponding momentum, considered as a function of x, and the circle around the sign of integral simply means that we integrate the momentum all the way along the closed phase trajectory. This will indeed give us the area inside it.

Let us start integration from the leftmost point and go first along the upper branch of the trajectory in Figure 1.8, from x_{min} to x_{max}. By its very definition, the integral $\int_{x_{min}}^{x_{max}} p_x(x)dx$ gives the area of the upper half of our curve. Now we can integrate along the lower branch of the curve, from x_{max} to x_{min}. Since here the coordinate changes from the greater to the smaller values ($dx<0$), the integral will acquire the negative sign – but the momentum $p_x(x)$ is negative there too! Therefore the second half of the integral will produce the remaining part of the area with *positive* sign. (To come to this conclusion, we could simply rotate our drawing 180 degrees!) The two contributions add up to the total area inside the curve.

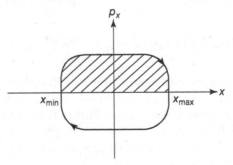

Figure 1.8 Action: the area enclosed by a phase trajectory

Hamilton, Hamiltonian, Hamilton's equations and state vectors

Now let us return to the energy, $H(x,p_x)$. It is denoted by H and is called Hamilton's function, or the *Hamiltonian*, to honour W.R. Hamilton (1805–1865). This great 19th century physicist and mathematician thoroughly reworked and expanded mechanics. His formulation of mechanics (*Hamiltonian mechanics*) not only revolutionized the classical physics of the time, but also turned out to be extremely important for the development of quantum mechanics almost a century later.

To put it simply, the Hamiltonian is the energy of a mechanical system expressed through its coordinates and momenta.

How will the harmonic oscillator's equation of motion look after we replace the velocity with the momentum? Since

$$p_x = mv_x = m\frac{dx}{dt},$$

then

$$\frac{d^2x}{dt^2} = \frac{d}{dt}\left(\frac{dx}{dt}\right) = \frac{d}{dt}\left(\frac{p_x}{m}\right) = -\omega_0^2 x.$$

Therefore now we get a set of two equations:

$$\frac{dx}{dt} = \frac{p_x}{m} \qquad \frac{dp_x}{dt} = -m\omega_0^2 x.$$

So far we have simply transformed *one* differential equation of the *second* order (that is, containing the second derivative over time) for *one* unknown function $x(t)$ into a system of *two* differential equations of the *first* order (containing only the first derivative over time) for *two* unknown functions $x(t)$ and $p_x(t)$.

This already brings some advantage, since the position and momentum to be taken seriously as the coordinates of the phase space should be considered as *independent variables* (like x, y and z in our conventional space). Now look at the Hamiltonian of the harmonic oscillator and notice that

$$m\omega_0^2 x = \frac{\partial}{\partial x}H(x, p_x), \quad \text{and} \quad \frac{p_x}{m} = \frac{\partial}{\partial p_x}H(x, p_x).$$

The cursive ∂'s denote *partial derivatives* – that is, the derivatives with respect to one independent variable, while all the others are fixed.

So, finally we can rewrite the equations of motion for the harmonic oscillator as

$$\frac{dx}{dt} = \frac{\partial}{\partial p_x}H(x, p_x);$$

$$\frac{dp_x}{dt} = -\frac{\partial}{\partial x} H(x, p_x).$$

These are the famous *Hamilton's equations*, which hold for any mechanical system. This is a set of two first order differential equations for two *conjugate variables* – a coordinate and a momentum.

Key idea: The partial derivative

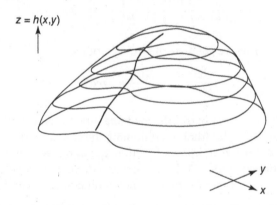

$z = h(x,y)$

The height of a hill, h, is a function of our position. If we point coordinate axes north (Oy) and east (Ox), then the incline we meet when travelling due north is precisely the *partial derivative*:

$$\frac{\partial}{\partial y} h(x, y) \equiv \lim_{\Delta y \to 0} \frac{h(x, y + \Delta y) - h(x, y)}{\Delta y}.$$

For a more complex system – say, a set of N particles in three-dimensional space, which is a good idealization for a gas of monoatomic molecules – the only difference is that we introduce more variables, such as $(x_1, p_{x1}, y_1, p_{y1}, z_1, p_{z1}, ..., x_N, p_{xN}, y_N, p_{yN}, z_N, p_{zN})$; – $3N$ components of radius vectors and $3N$ components of the momenta, and write $6N$ Hamilton's equations instead of two. Each couple of conjugate variables describe a *degree of freedom* – thus a one-dimensional harmonic oscillator has one degree of freedom, and a set of N material points in a three-dimensional space have $3N$ degrees of freedom.

We can take the set of coordinates and momenta and consider them as the coordinates of a single point in a $6N$-dimensional phase space – or, if you wish, as components of a radius vector in this space (Figure 1.9):

$$\boldsymbol{R} \equiv (x_1, p_{x1}, y_1, p_{y1}, z_1, p_{z1}, \ldots, x_N, p_{xN}, y_N, p_{yN}, z_N, p_{zN}).$$

One can call it the *state vector*, since it completely determines both the *current state* of the system and its *future evolution*, governed by the Hamilton's equations.

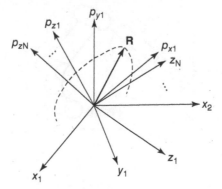

Figure 1.9 *6N*-dimensional phase space of a system of *N* particles

Generalized coordinates, canonical momenta, and configuration space

Now as we emerge from the thicket of mathematical symbols burdened by a load of new symbols, equations and definitions, we can ask ourselves: was it worth it? What did we learn about a harmonic oscillator that was not already clear from the Newton's second law? To be honest, no – we did not need all this fancy equipment just to treat a harmonic oscillator. Neither did Hamilton. His reasons for developing the Hamiltonian mechanics were nevertheless very practical.

Newtonian mechanics are well and good, when one deals with one or several point-like objects moving freely in space, such as the Sun and the planets (even though these problems are in no way trivial). Now suppose you have a complex mechanism like one of those contraptions the Victorians were so adept at building (Figure 1.10) and need to predict its behaviour. Describing everything in terms of positions of different parts of the mechanism in real space would be incredibly messy and impractical. For example, why give the positions and velocities in the real space of the ends of a rigid lever in 12 numbers, which can be completely described by just two numbers – a rotation angle and the corresponding angular velocity? But on the other hand, we do need positions, velocities and accelerations in the real space in order to write down the second law!

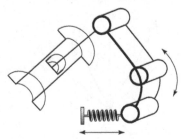

Figure 1.10 A mechanical contraption, which is hard to describe by Cartesian coordinates

This difficulty became clear before Hamilton's time, and was eventually resolved by another great scientist, Lagrange. He developed a very useful and insightful approach – *Lagrangian mechanics* – that served as the basis for the Hamiltonian one, which is very much in use and also very important for the modern quantum physics. We will discuss this further in Chapter 9.

Spotlight: Lagrange equations

Lagrange equations – just to impress you. We will only need them in Chapter 9.

$$\frac{d}{dt}\left(\frac{\partial L(q,\dot{q})}{\partial \dot{q}}\right) - \frac{\partial L(q,\dot{q})}{\partial q} = 0.$$

The gist of it is that Lagrange developed a method of describing an arbitrary mechanical system by a set of *generalized coordinates*, q_i and *generalized velocities*, $\dot{q}_i \equiv d\dot{q}_i/dt$. These coordinates can be *any* convenient parameters that describe our mechanical system: rotation angles; elongations; distances; or even conventional positions in real space. The index $i = 1, 2, ..., M$ labels these coordinates; M is the number of the degrees of freedom our system possesses. Please note that, by tradition and for brevity, for the generalized velocities the Newtonian notation for a time derivative is used.

Key idea: Phase and configuration spaces

The $6N$ variables $(x_1, p_{x1}, y_1, z_1, p_{z1}, ...$
$..., x_N, p_{xN}, y_N, p_{yN}, z_N, p_{zN})$ which completely describe a set of N particles in three-dimensional space, can be thought of as coordinates of a single particle in a $6N$-dimensional phase space of the system. If we are only interested in the positions of these particles, we can also describe the system as a single point with coordinates $(x_1, y_1, z_1, ..., x_N, y_N, z_N)$, which moves in a $3N$-dimensional *configuration space*. In either case, our investigation of a totally mundane system of classical particles in a conventional three-dimensional space takes us to pretty esoteric spaces.

Lagrange discovered how to write a certain function – the *Lagrangian*, $L(q, \dot{q})$ – of all these variables, which satisfies certain differential equations, from which all the generalized coordinates $q_i(t)$ can be found for all later moments of time, provided that we specify the initial conditions: $q_i(0) = q_{0i}$; $\dot{q}_i(0) = \dot{q}_{0i}$. Moreover, he proved that these equations are equivalent to those following directly from Newton's second law, even though the former are much easier to write down than the latter.

As before, we can take the set of all generalized coordinates and consider them as the coordinates of a single point in an M-dimensional configuration space. And, as before, we can introduce the phase space (which will be, of course, $2M$-dimensional), the state vector in this space, and the

Hamilton's equations. That is, Hamilton did it. In particular, he discovered how – starting from a Lagrangian – one can find the momentum p_i, corresponding to a given generalized coordinate q_i. It is called the *canonical momentum* and is often something very unlike $m_i\dot{q}_i$. Then he showed how to obtain from the Lagrangian $L(q,\dot{q})$ the Hamiltonian $H(q,p)$. And then one can immediately write down and – with some luck and much hard work – solve the Hamilton's equations and find the behaviour of the system:

$$\frac{dq_i}{dt} = \frac{\partial H}{\partial p_i};$$

$$\frac{dp_i}{dt} = -\frac{\partial H}{\partial q_i};$$

$$i = 1, 2, \dots 2M$$

In many cases these equations are easier to solve than the Lagrange equations. What is more important to us is that they form a vital link between quantum and classical mechanics.

Hamilton's equations describe the evolution with time of the state vector in a $2M$-dimensional phase space of our system. One cannot easily imagine this, but some geometrical intuition based on our lower-dimensional experience can be actually developed with practice. For example, you can draw the phase trajectories in each of the M planes (q_i, p_i) and consider them in the same way one does different projections of a complex piece of machinery in a technical drawing.

We will try to show how it works on another example – the last in this chapter – of those notions from classical physics, which turn out to be not that much different from their quantum counterparts.

Statistical mechanics, statistical ensemble and probability distribution function

One of the simplest and most useful model systems considered by classical physics is the *monoatomic ideal gas* – a collection of a *very* large number N of identical material points, which do not interact with each other. This model helped establish the fundamentals of *statistical mechanics* – the aspect of physics that explains, in particular, why water freezes and how an electric heater works. We already know that the state of this system can be completely described by a state vector in a $6N$-dimensional phase space (or, which is the same, a single point in this space – the end of the state vector; see Figure 1.9).

As we know, the state vector determines *all* the positions (in real space) and *all* the momenta of *all* N gas particles. A single cubic centimetre of gas at normal pressure and zero degrees Celsius contains about 2.7×10^{19} molecules. Therefore the state vector contains a lot of unnecessary and unusable information. It is unusable, because even if we had all this information, we could never solve $6 \times 2.7 \times 10^{19}$ Hamilton's equations per cubic centimetre of gas. Unnecessary, because we do not care where precisely each gas particle is placed and what momentum it has at a given moment of time. We are interested in bulk properties, like temperature, pressure, local flow velocity and so on at a given position in a given moment of time.

The central point of statistical mechanics makes it so that a single set of such bulk properties $(P(r, t), T(r, t), v(r, t), ...)$ – a so-called *macroscopic state* of our system – corresponds to a huge number of slightly different *microscopic states* (each described by a state vector in the $6N$-dimensional phase space). These are found in the vicinity of certain point $R = (q, p)$ in this space (Figure 1.11), which has 'volume' $\Delta\Gamma(q, p)$.

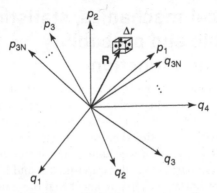

Figure 1.11 A statistical ensemble in phase space

One way to use this idea is to introduce a *statistical ensemble* of N-particle systems – that is, a very large number N_E of copies of our system with slightly different initial conditions. They will form something like a 'cloud of gas' in the phase space, and each will follow its own unique trajectory determined by the Hamilton's equations. (The fact that in practice we cannot solve or even write down these equations does not matter in the least!) To calculate the likelihood of finding our system in a given macroscopic state, we can simply count how many of N_E systems in the ensemble are found in the 'volume' $\Delta\Gamma(q, p)$ at a given moment. Now denote this number $\Delta N_E(q, p, t)$ and introduce the function

$$f_N(q, p, t) = \lim_{N_E \to \infty} \frac{\Delta N_E(q, p, t)}{N_E}.$$

This function is called the *N-particle probability distribution* (or simply *distribution*) *function*. It gives the probability to find the system in a given *micro*scopic state (because it still depends on *all* the momenta and *all* the positions of *all* the particles). Obviously, this is not much of an improvement, since this is as detailed a description, as we had initially, and the equations of motion for this function are as many, as complex and as unsolvable as the Hamilton's equations for all N particles. But once we have introduced the distribution function, some drastic simplifications become possible. To begin with we can ask what the probability is for *one* gas particle to have a given position and momentum at a given time. Let us denote it as

$$f_1(q, p, t),$$

where now (q, p) are the position and momentum of a single particle in a phase space with a miserly six dimensions.

Since all gas particles are the same, the answer will do for any and for all of them. Moreover, properties such as temperature or pressure, which determine the *macro*scopic state of gas, can be calculated using only $f_1(q, p, t)$. Finally, the approximate equations, which determine the behaviour of $f_1(q, p, t)$, can be written down and solved easily enough.

Sometimes we would need to know the probability that one particle has the position and momentum (q_1, p_1), and other (q_2, p_2), simultaneously. The approximate equations for the corresponding two-particle distribution function, $f_2(q, p, t)$, can be also readily written down and – with somewhat more trouble – solved. Such equations can be written for all distribution functions, up to the N-particle one from which we started, but in practice physicists very rarely need any distribution functions beyond $f_1(q, p, t)$ and $f_2(q, p, t)$.

The idea that sometimes we can and need to know something only on average is a very powerful idea.[4] It allowed the great progress of classical statistical mechanics. We will see how it is being used in quantum mechanics.

---

4   In statistical mechanics, knowledge is power, but ignorance (of unnecessary data) is might!

Liouville equation*

One last remark: the N-particle distribution function $f_N(q, p, t)$ changes with time, because the coordinates and momenta also change with time. One can derive the equation for $f_N(q, p, t)$:

$$\frac{df_N(q,p,t)}{dt} = \frac{\partial f_N}{\partial t} + \sum_j \left\{ \frac{\partial f_N}{\partial q_j} \frac{dq_j}{dt} + \frac{\partial f_N}{\partial p_j} \frac{dp_j}{dt} \right\} = 0,$$

which is essentially the result of applying the chain rule of differentiation to the time derivative of $f_N(q, p, t)$. This equation describes the 'flow' in the phase space of the points, which represent the individual systems in the statistical ensemble. Since the *total* number of systems in the ensemble is constant, the *full* time derivative of $f_N(q, p, t)$ must be zero. For the *partial* time derivative, which describes how $f_N(q, p, t)$ changes in time at a *fixed* point (q, p) in the phase space, we therefore obtain the *Liouville equation*,

$$\frac{\partial f_N(q,p,t)}{\partial t} = - \sum_j \left\{ \frac{\partial f_N}{\partial q_j} \frac{dq_j}{dt} + \frac{\partial f_N}{\partial p_j} \frac{dp_j}{dt} \right\}.$$

Fact-check

1 Velocity is
   a $d^2\mathbf{r}/d^2t$
   b $d\mathbf{r}/dt$
   c $dt/d\mathbf{r}$
   d $d^2\mathbf{r}/dt^2$

2 Acceleration is
   a $d^2\mathbf{r}/d^2t$
   b $d\mathbf{r}/dt$
   c $dt/d\mathbf{r}$
   d $d^2\mathbf{r}/dt^2$

3 In order to predict the behaviour of a material point you need to know its initial

   a   position
   b   position and velocity
   c   velocity and acceleration
   d   position and momentum

4 How many dimensions does the phase space of a system with nine spring pendulums have?

   a   3
   b   9
   c   54
   d   18

5 How many dimensions does the configuration space of the same system have?

   a   3
   b   9
   c   54
   d   18

6 The phase trajectory of a harmonic oscillator can have the following shape:

   a   parabola
   b   hyperbola
   c   ellipse
   d   circle

7 Indicate impossible trajectories of a material point in real space

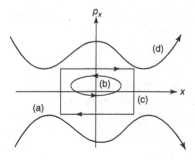

8 Indicate possible trajectories of a material point in the phase space

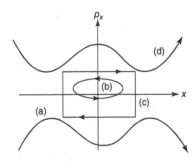

9 Conjugate variables are

   a x and p_y

   b p_x and p_z

   c z and p_z

   d x and y

10 Hamilton's equations directly describe the evolution of

   a   mechanical system in real space

   b   state vector in phase space

   c   state vector in configuration space

   d   only of a harmonic oscillator

Dig deeper

Further reading I. Some serious reading on classical mechanics and related mathematics.

R. Feynman, R. Leighton and M. Sands, *Feynman Lectures on Physics*, Vols I, II and III. Basic Books, 2010 (there are earlier editions). See in particular Vol. I (read there about mechanics and some mathematics, Chapters 1–25).

P. Hamill, *A Student's Guide to Lagrangians and Hamiltonians*. Cambridge University Press, 2013.

F. Reif, *Statistical Physics*, Berkeley Physics Course, Vol. 5. McGraw-Hill Book Company, 1967.

L. Susskind and G. Hrabovsky, *Theoretical Minimum: What You Need to Know to Start Doing Physics*. Basic Books, 2013. Here is a really welcome book: classical mechanics, including its more arcane sides (like phase space and Hamilton equations), presented in an accessible form! Susskind is a prominent string theorist, which makes it even more interesting.

Ya. B. Zeldovich and I. M. Yaglom, *Higher Math for Beginning Physicists and Engineers*. Prentice Hall, 1988. Introduces mathematical tools together with the physical problems they were developed to solve, and teaches learners how to use these tools in practice.

Further reading II. About the ideas and scientists.

V.I. Arnold, *Huygens & Barrow, Newton & Hooke*. Birkhauser, 1990. The same story, told by a great mathematician from his unique point of view, and much more.

J. Bardi, *The Calculus Wars*. High Stakes Publishing, 2007. How Newton and Leibniz discovered calculus.

L. Jardine, *The Curious Life of Robert Hooke: The Man who Measured London*. Harper Perennial, 2004. A very good biography.

T. Levenson, *Newton and the Counterfeiter*. Faber & Faber, 2010. A fascinating story of how Newton managed the Royal Mint, tracked and caught a gang of counterfeiters, and had their chief executed.

2

Less familiar physics in stranger spaces

The objective of the previous chapter was to glance at classical mechanics from a less familiar angle.

Now we will try and look at quantum mechanics from this angle, in the hope that it will look somewhat less strange – or at least no stranger than it must. This is not how it was seen by its founders, but it is only natural: highways rarely follow the trails of the first explorers.

Let us start from the beginning. In 1900, Professor Max Planck of the University of Berlin introduced the now famous *Planck constant* $h = 6.626\ldots\cdot10^{-34}$ J s. For very compelling reasons, which we will not outline here, Max Planck made the following drastic prediction concerning harmonic oscillators, which in modern terms went as follows:

A harmonic oscillator with frequency ν_0 can change its energy only by an amount $h\nu_0$ (the so-called energy *quantum*, from the Latin *quantus* meaning 'how much').

In other words, $\Delta E = h\nu_o$.

Initially Planck referred to what we now call 'energy quantum' of an oscillator as one of 'finite equal parts' of vibrational energy. The term 'quantum' was coined later. One of its first users, along with Planck, was Philipp Lenard, a first-rate experimentalist, whose work greatly contributed to the discovery of quantum mechanics, and who was also – regrettably – a first-rate example of how a brilliant physicist can go terribly wrong outside his field of expertise. But, no matter what terms were used, the notion of deriving discrete portions of energy from an oscillator was revolutionary.

Key idea: Planck constant

In modern physics literature, the original Planck constant $h = 6.2606957(29)\bullet10^{-34}$ J s is used less frequently than $\hbar$ (pronounced 'eichbar'):

$$\hbar = \frac{h}{2\pi} = 1.054571726(47)\cdot10^{-34}\,\text{J s}.$$

Both are usually called 'the Planck constant'.

The figures in brackets indicate the accuracy to which this constant is currently known. Roughly speaking, if some $A = 1.23(45)$, it means that A is most likely equal to 1.23, but can be as small as $1.23 - 0.45 = 0.78$ or as large as $1.23 + 0.45 = 1.68$.

Spotlight: Quantum vs quantus

Why do we have *quantum* mechanics and not *quantus* mechanics? In Latin, as in German, Russian, and many other languages, words undergo some serious declination, and *quantus* becomes *quantum* in such expressions as 'quantum satis' (the amount which is needed). The plural, *quanta*, is also a Roman legacy.

Key idea: Quantization condition

Please note one important subtlety. The quantization condition does not actually say that the energy of an oscillator must be a multiple of $h\nu_0$, but only that it can only *change* by this amount. That is, its energy is

$$E = (n - \alpha)h\nu_0$$

where $n = 0,1,2,...$, and α can be any number between zero and one. Of course, at the moment everybody, including Planck himself, assumed that $\alpha = 0$. Later in 1913, Einstein and Stern found that for a harmonic oscillator $\alpha = 1/2$. This does make a difference, and we will return to this important point a little later.

Indeed, an oscillator is, if you like, a pendulum. The energy of its oscillations depends on how far we swing it from the vertical at the beginning – or how strongly we have pushed it. The idea that one cannot push a pendulum with an arbitrary force sounded as strange to a contemporary physicist as if somebody had told him that he could not walk, but could only jump by exactly equal bounds in rigidly determined directions, like a chess piece. Nevertheless, a closer look at this strange situation shows that it is not as strange as it seems – but even stranger.

Quantum oscillator and adiabatic invariants

Let us return to the phase space picture of a classical harmonic oscillator (Figure 2.1). Its phase trajectory is an ellipse with area

$$S = \frac{2\pi}{\omega_0} H\left(x_o, p_{xo}\right) = T_o H\left(x_o, p_{xo}\right),$$

where $H(x_0, p_{x0}) = E$ is the oscillator energy. What consequences will follow from the *energy quantization condition*, $\Delta E = h\nu_0$, or, as it is usually written in modern books and articles, $\Delta E = \hbar\omega_0$?

There seems to be little to worry about when dealing with the events in our everyday life. Planck constant is exceedingly small. Take as an example of an oscillator a child's swing, with a typical frequency of about one swing per second, which corresponds to the linear frequency of 1 Hz. Then the corresponding energy quantum is $h \cdot 1$ Hz $= 6.626... \cdot 10^{-34}$ J. This is a very small amount of energy indeed. An average ant (which weighs about 0.5 mg) falling from a height of 1 mm will acquire energy about 10^{25} times greater than that. For comparison, the Sun radiates approximately 10^{25} times more energy per second than a standard 40 W light bulb (and remember here we count all of the Sun's energy production, not just the tiny amount that reaches our insignificant planet). So, at least from the point of view of everyday physical activities, quantization of energy does not seem to lead to any perceptible effects.

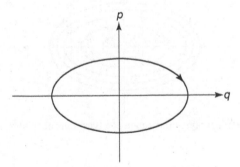

Figure 2.1 Phase trajectory of a harmonic oscillator

All the same, the quantization condition means that an oscillator with frequency ω_0 cannot follow an *arbitrary* phase trajectory:

$$S_n = T_0 \cdot (n + \alpha)\hbar\omega_0 = \frac{2\pi}{\omega_0}(n + \alpha)\hbar\omega_0 = 2\pi \cdot (n + \alpha)\hbar$$

$$= (n + \alpha)h; \; n = 0, 1, 2, ...; \; \alpha = \frac{1}{2}.$$

Only the ellipses with strictly defined areas are allowed! (Figure 2.2).

This is a remarkable fact. The oscillator energy quantization was not universal: different oscillators would have different frequencies and therefore different energy quanta. Returning to our earlier example, a gnat, which is annoying the child on a swing with its wings beating at 20,000 times per second, can change their energy by portions, which are 20,000 times greater – though still imperceptibly small. But the area enclosed by the phase trajectory of *any* harmonic oscillator, no matter its frequency, is always given by a multiple of the original Planck constant h:

$$S_n = (n+\alpha)h; \quad n = 0, 1, 2, ...; \quad \alpha = \frac{1}{2}.$$

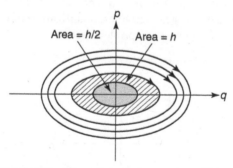

Figure 2.2 Quantization of the action of a harmonic oscillator

You recall that this area is called *action* and can be written as $S = \oint p_x(x)dx$. Therefore, the Planck constant h is called the *quantum of action*.

The relation

$$S = \oint p_x(x)dx = (n+\alpha)h; n = 0,1,2.\ ...;0 \le \alpha < 1,$$

is known as the *Bohr–Sommerfeld quantization condition,* and it holds not just for a harmonic oscillator; it holds true in every case, when such an integral can be calculated, e.g. for a non-

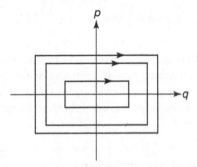

Figure 2.3 Quantization of the action of a non-harmonic oscillator

harmonic oscillator (Figure 2.3), but the value of α may depend on the system.[1]

Spotlight: Philipp Lenard

Philipp Lenard (1862–1947) was a prominent German experimental physicist and Nobel Prize winner (1905). He made a number of important contributions to physics, the foremost being his research on cathode rays (that is, electron beams – but that was before the very existence of electrons was discovered) and photoelectric effect. The latter work was one of the experimental foundations on which quantum theory was built: in one of his famous papers of 1905, Einstein explained Lenard's experiments using the idea of energy quantization. Sadly, between the wars Lenard became a Nazi sympathizer and spent the later part of his life promoting 'German physics' and fighting the 'unGerman' one.

Here we run into a problem. Let us consider, for example, a pendulum. Its frequency ω_0 depends on the length of the string

---

1   This became clear later on – initially everyone assumed that $\alpha = 0$.

and on the gravity acceleration, and if the oscillation amplitude is small, $\omega_0 = \sqrt{g/l}$. Assuming that Planck, Einstein and Stern are right, the energy of this system must be

$$E = \left(n + \frac{1}{2}\right)\hbar\omega_0 = \left(n + \frac{1}{2}\right)h\nu_0.$$

But we can always slightly shorten or lengthen the string and thus change the frequency by any amount whatsoever. How does this agree with our earlier insistence that energy can only change by finite steps – even if these steps are tiny by our everyday measure?

Key idea: Adiabatic invariant

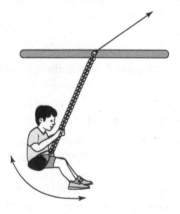

An *adiabatic invariant* of a physical system is a physical quantity, which stays constant, if some parameter of this system is being changed *very* slowly. (Such infinitesimally slow change is called adiabatic. This word is used in a different sense in some other branches of physics.)

The ratio of the frequency of a harmonic oscillator to the energy of its oscillations is an adiabatic invariant. For example, if you take a pendulum and very slowly pull the string (or let it go), the frequency and the amplitude of oscillations will change – but in such a way as to keep this ratio constant.

Spotlight: Paul Ehrenfest

Paul Ehrenfest (1880-1933) made deep and lasting contributions to theoretical physics. One of his (quite unjustified) worries was that he was not as good a physicist as his best friends – Albert Einstein and Niels Bohr.

Paul Ehrenfest noted that the *ratio E/v₀* does not change if the frequency of the oscillator is altered *infinitesimally slowly* – or, using a shorter term, *adiabatically*. Then, even though the energy of the system changes, the number n of its quanta stays the same!

$$n + \frac{1}{2} = \frac{E}{\hbar\omega_o} = \frac{E}{hv_o} = const.$$

They say that the ratio E/v_o is an *adiabatic invariant*. What is more, it is known from classical mechanics that the area inside the phase trajectory, that is, the integral $\oint p_x(x)dx$, is always an adiabatic invariant for any physical system for which it exists. This area, as we remember, is the action. So what Ehrenfest actually says is that you can quantize action using the Bohr–Sommerfeld quantization condition and will not run into any problems.[2]

Action quantization, phase space and the uncertainty principle

In Figure 2.3 we see the phase space of an oscillator sliced into layers of area h due to the Bohr–Sommerfeld quantization condition. Is it possible to account for it in a more general case?

---

2   Well, actually you will – because not every system has nice closed phase trajectories. Fortunately those were not important enough and did not have to be considered when the very basics of quantum mechanics were being discovered. Modern quantum mechanics has developed tools for dealing with them.

Let us take some one-dimensional system, which will have, of course, a two-dimensional phase space. Considering this space just as a usual plane from elementary geometry, we could draw any number of figures *but* due to the quantization, the area of any such figure will be a multiple of h. We could say that the unit element of the phase space of a quantum system has the area

$$\Delta S = \Delta x \cdot \Delta p = h.$$

What about more complex systems? We know that, for example, a system of N particles can be described by a single vector $\boldsymbol{R} \equiv (x_1, p_{x1}, y_1, p_{y1}, z_1, p_{z1}, ..., x_N, p_{xN}, y_N, p_{yN}, z_N, p_{zN})$ in a $6N$-dimensional phase space. Well, now instead of an 'area' we should talk about a 'volume' – or rather 'hypervolume'. A unit element of such a phase space will have the hypervolume (Figure 2.4).

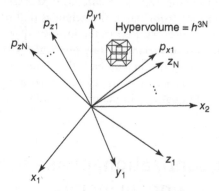

Figure 2.4 Elementary hypervolume of a *6N*-dimensional phase space

$$\Delta V = (\Delta x_1\, \Delta p_{x1})\,(\Delta y_1\, \Delta p_{y1})\,(\Delta z_1\, \Delta p_{z1})\cdots(\Delta x_{Nn}\, \Delta p_{xN})\,(\Delta y_N\, \Delta p_{yN})\,(\Delta z_N\, \Delta p_{zN}) = h^{3N}$$

Apart from this trivial difference, the situation remains the same: the action *for each pair of conjugate variables* (e.g. y_1, p_{y1}) can be only changed by an amount nh.

This is one manifestation of a fundamental property of Nature: the *quantum uncertainty principle*. In its 'folk' form it states:

The product of uncertainties of two conjugate variables cannot be less than the Planck constant (actually, half the Planck constant).

Conjugate variables are those related by the Hamilton equations and their dimensionalities are as such to make the dimensionality of their product the same as that of the Planck constant, J·s. There are lots of pairs of conjugate variables in physics, and we will meet some of them later – but the position and momentum are the most important ones for our current purpose.

On the surface of it, the uncertainty principle sounds pretty simple; for some reason we cannot know exactly the position and the momentum (i.e. velocity) of a particle. So what? Suppose the phase space is neatly sliced in squares of area h (or 'hypercubes' of volume h^{3N}). If that were the case, the Universe would be something like a chessboard with very small squares, and with game pieces – elementary particles – making very small jumps between them (which would, probably, require slicing time in tiny bits Δt as well). Of course then we would only know the position and momentum of each piece up to the size of the square it occupies and would not care, since a rook can take, say, a bishop no matter where in its square it stands (Figure 2.5).

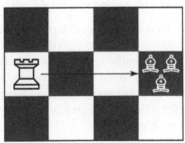

Figure 2.5 'Quantum chess', or how Nature does NOT play

Such a world picture would profoundly shock a Planck's contemporary, accustomed to the mechanics of continuum and infinitely divisible space and time, but not so much a modern person acquainted with computer graphics. And it would be a very simple picture indeed. Instead of Hamilton's *differential* equations for *continuous* variables x and p, we would have *difference* equations for *discrete* variables $x_n = n\Delta x$ and $p_m = m\Delta p_x$, something like

$$\frac{\Delta x_n}{\Delta t} = \frac{H(x_n, p_{m+1}) - H(x_n, p_m)}{\Delta p_x};$$

$$\frac{\Delta p_m}{\Delta t} = -\frac{H(x_{n+1}, p_m) - H(x_n, p_m)}{\Delta x}.$$

Actually, this is how differential equations *are* being solved numerically from the times of Sir Isaac Newton. Some refinements developed since do not affect the basic idea. These numerical solutions are the basis for all modern technology. In particular, they provide very accurate descriptions of how aeroplanes fly, how cars run, how steel beams support skyscrapers and steel cables hold up bridges, and how space apparatus approach other planets. In all of these cases even the smallest value of the product $\Delta x \Delta p_x$ is much, much greater than the Planck constant. If quantization simply reduced to the 'chessboard' picture, Nature would have been such a straightforward and easy to understand thing!

Unfortunately, we have already seen that this is not so, already in the case of a linear oscillator; the phase trajectories are elliptic and the distances between the consecutive allowed trajectories grow smaller and smaller as their size increases. The Bohr–Sommerfeld condition only means that any *change* in the action of a physical system – that is, the area between two of its phase trajectories – is a multiple of h.

How should we describe our system to accommodate such behaviour?

Zero-point energy and complex numbers

Let us look at the situation from a different angle. A phase trajectory is traced by the end of the state vector *R(t)*. Suppose the oscillator is at rest. Then the trajectory reduces to the single point at the origin, and the energy of such an oscillator is zero.

This is well and good for a classical oscillator. But a quantum oscillator has the minimum energy (so-called *zero-point energy*) $E_0 = \left(\frac{1}{2}\right)\hbar\omega_0 = \left(\frac{1}{2}\right)h\nu_0$ and the minimum action $S_0 = \left(\frac{1}{2}\right)h$. This is because of the value of $\alpha = \frac{1}{2}$ in the Bohr–Sommerfeld quantization condition for the oscillator, found by Einstein and Stern. The phase trajectory for a quantum oscillator 'at rest' is therefore not a *point* at origin, but an *ellipse with area h/2* – and this is the smallest area that can be enclosed by a phase trajectory. What is then the meaning of a point in the phase space?

Key idea: Zero-point energy

Zero-point energy of a linear oscillator with frequency $\omega = 2\pi\nu$ is the minimal energy it can have. It is equal to

$$E_0 = \frac{h\nu}{2} = \frac{\hbar\omega}{2}.$$

Zero-point energy is a manifestation of the fundamental uncertainty principle; if the oscillator were at rest, both its position and momentum would have precise values, i.e. zero. The existence of zero-point energy is confirmed by numerous experiments, and is, e.g. behind strange properties of liquid helium.

This is a question of fundamental importance. Indeed, if even at rest a quantum oscillator can only be 'located' within an area h/2 of the phase space, the very concept of a 'point' – that is, an oscillator having simultaneously a position x and a momentum p_x – becomes tenuous. But is this so important? After all, h is *very* small. Can we not just accept that a 'point' in the phase space is actually a dot of area h/2, all the phase trajectories have a finite thickness of order $\sqrt{h}$, and be done with it? Surely, at least when we describe macroscopic systems – stars in space, children on swings, ants falling off a blade of grass, bacteria swimming in drops of water – any system, whose action is much greater than h, this would not make any difference?

Case study: Complex matters

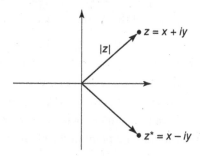

Complex numbers can be represented as points in a plane (*complex plane*). A point, representing the complex number

$$z = x + iy$$

has the coordinates x along the *real axis* and y along the *imaginary axis*.

A *complex conjugate* number to a number z is denoted by a star and equals

$$z^* = x - iy.$$

The distance from the origin to the point z is denoted by $|z|$ and, of course, equals

$$|z| = \sqrt{\{x^2 + y^2\}}.$$

Since $i^2 = -1$, you can see that

$$|z|^2 = (x + iy)(x - iy) = z \cdot z^*.$$

$|z|$ is called the *absolute value*, or the *modulus*, of z, and it is always a normal, *real* number.

The correct answer is that it depends. It was actually a big surprise, when – quite recently – it turned out that pretty large (macroscopic) systems can and do behave in the same way as microscopic particles (like electrons), that is, quantum mechanically, no matter how big their action is compared to h.[3] But special measures must be taken to observe such a behaviour, and in the vast majority of cases, large systems (i.e. those with $S \gg h$) indeed do behave simply as if their phase trajectories were drawn by a pen of thickness $\sqrt{h}$ per a degree of freedom – and this does not make any difference.

Key idea: Euler's identity

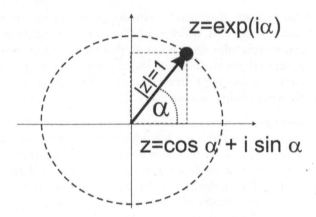

The exponent of a purely imaginary number $i\alpha$ is given by

$$z = e^{i\alpha} = (\cos \alpha + i \sin \alpha).$$

The point representing the complex number z has thus coordinates $x = \cos \alpha$ and $y = \sin \alpha$, and α can be considered as the angle between the real axis and $0z$.

It is called the *argument*, or the *phase* of the complex number z. Take $\alpha = \pi$ (we measure angles in the scientific fashion, using radians instead of degrees: π radians equal 180°). Then

$$e^{i\pi} = (\cos \pi + i \sin \pi) = -1.$$

This is *Euler's identity*, which is considered the most beautiful mathematical formula ever.

---

3 See Chapters 11, 12.

But this clearly cannot be so for small, microscopic systems, whose action is only few times the Planck constant. Moreover, in any case we get into an unpleasant logical loop – what is properly known in Latin as *circulus vitious*. Indeed, due to the Bohr–Sommerfeld quantization we cannot with certainty talk about a *point* in the phase space. Then the very concept of phase *trajectories* becomes rather vague, which undermines the Bohr–Sommerfeld quantization condition since it relies on the existence of phase trajectories, which are infinitesimally thin.

All of this indicates that there is something wrong with our neat phase space description, which worked so nicely in the classical case. Can we relax this description, so that we would not have to pinpoint *both* the position and the momentum of a particle, and ensure that the quantization condition and the uncertainty principle would hold automatically?

The answer is, of course, yes. Moreover, the new description is not quite unlike the old one – but with some weird twists. One of these twists is that we will have to use *complex numbers*. These numbers are a very convenient instrument of dealing with many problems in mathematics, physics, engineering and even economics. Each complex number can be written as a sum:

$$z = x + iy.$$

Here x is called its *real*, and y its *imaginary* part – both x and y are usual, *real* numbers, while i is the *imaginary unit*.[4] Apart from a strange, but justified property, that

$$i^2 = -1,$$

complex numbers do not present any trouble. For example, one can add, subtract, multiply and divide them as usual numbers, extract roots, calculate arbitrary functions of them, etc. It

---

4   The terms 'real' and 'imaginary' come from earlier times when mathematicians were less acquainted with strange and unconventional mathematical objects and were not sure how to deal with them. To think of it, they considered all numbers, which cannot be expressed as normal fractions, *irrational*, and some of them even *transcendent*! By the time these objects became well understood, it was pointless to rename them.

only takes some practice. For example, $(1 + 2i)(3 - 4i) = 3 + 6i - 4i - 8i^2 = 3 + 8 + (6 - 4)i = 11 + 2i$. Instead, they allow simplification of many mathematical operations.

One of the most useful simplifications one can get from complex numbers is the *Euler's formula* for the exponent of a complex number,

$$e^z = e^{x+iy} = e^x (\cos y + i \sin y).$$

Spotlight: Leonhard Euler

Leonhard Euler (1707–1783) was a great Swiss mathematician, who made revolutionary contributions to mathematics and physics. Calculus as we use it now is to a significant degree shaped by Euler.

Euler spent most of his extremely productive life in St Petersburg and in Berlin. When he died, Condorcet wrote: 'il cessa de calculer et de vivre' (he ceased to calculate and to live).[5]

Why is it a simplification? As you know from trigonometry, the formulas for adding sines and cosines, or finding sines and cosines of sums of arguments, are messy. On the other hand, it is simple to find a sum or a product of exponentials of any numbers – complex or not. Extracting the results for the trigonometric functions is then a pretty straightforward task.

State vector and Hilbert space

Now let us return to the quantum description of a single particle in one dimension.

Since we cannot pinpoint both its position and momentum, let us deal with the position only. The state of the particle will be described by a quantum *state vector*. We will denote it by $\psi(x)$. This is a function of the position x, and we assume that it can take complex values.

5 'Eloge de M. Euler', *History of the Royal Academy of Sciences* (Paris, 1786): 68.

By calling $\psi(x)$ a vector, we imply that it should indicate a direction in some space. What kind of space may it be? This cannot be a usual space. In a usual space a vector will simply point at the position of a particle, we would have to supplement it with the instantaneous velocity of the particle, and this would bring us back to the classical mechanics of Chapter 1. The space we will be dealing with is called the *Hilbert space*, after David Hilbert, who discovered them.

Hilbert spaces are spaces *of functions*. What does it mean? Let us consider an example: a string of length L with fixed ends. Assume that it is infinitely stretchable and elastic and can be put in any shape $f(x)$ (Figure 2.6).

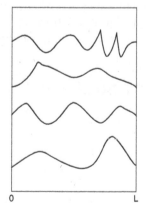

Figure 2.6 Each shape of a string can be regarded as a *point* in a Hilbert space

It is known that any such function $f(x)$ can be represented as a trigonometric Fourier series, i.e. a sum of sine and cosine terms:

$$f(x) = \sum_{n=0}^{\infty} A_n \cos\frac{2\pi n}{L}x + \sum_{n=1}^{n} B_n \sin\frac{2\pi n}{L}x$$

Musically, this is just the representation of a complex chord in terms of simple harmonic tones. Each of the 'tones' is independent from the rest of them. Mathematically this means that if you integrate a product of two different 'tones', you obtain zero:

$$\int_0^L dx \, \sin\frac{2\pi n}{L}x \cdot \sin\frac{2\pi m}{L}x = 0 \text{ unless } n = m;$$

$$\int_0^L dx \, \cos\frac{2\pi n}{L}x \cdot \cos\frac{2\pi m}{L}x = 0 \text{ unless } n = m.$$

$$\int_0^L dx \, \sin\frac{2\pi n}{L}x \cdot \cos\frac{2\pi m}{L}x = 0 \text{ for any } m, n.$$

Mathematicians are great at finding similarities where there seem to be none. So they found that the expansion of $f(x)$ is similar to the expansion of a conventional vector in an *orthogonal basis*:

$$A = \sum_{j=1}^{N} A_j e_j.$$

Here the unit vectors e_j are the basis vectors – in our conventional space they are simply vectors of unit length along the directions *Ox, Oy* and *Oz*, and they are all at right angles to each other, that is, their *scalar* (or *dot*) *product* is zero:

$$e_j \cdot e_k = 0 \text{ if } j \neq k.$$

Now, we know how to calculate a scalar product of two vectors in conventional space. But the mathematicians found that if you consider functions as vectors, you can consider their integral as a scalar product. To be more specific, for any two functions, say $f(x)$ and $g(x)$– any two 'shapes of the string' – we can define the 'scalar product' (which is actually called a scalar product – why invent new terms, when the existing ones will do?) as

$$\langle g \,|\, f \rangle = \int_0^L g^*(x)f(x)dx.$$

We are taking here a complex conjugate of the function $g(x)$, because, as you have been warned, we will need to use complex numbers. Of course, because of this the scalar product of two functions now depends on their order, unlike the conventional scalar product of two vectors:

$$\langle f|g\rangle = \langle g|f\rangle^*$$

In the same way as we express the length of a conventional vector through its scalar product with itself, $|A| = \sqrt{A \cdot A}$, we can define the 'length' of a function (it is called *norm* and is denoted by double lines to stress that we are not dealing with a usual vector) as the square root of the integral:

$$\|f(x)\| = \sqrt{\langle f|f\rangle} = \sqrt{\int_0^L |f(x)|^2\, dx}$$

This will be a conventional, real number. Any 'vector' $f(x)$ can be *normalized*, that is, multiplied by some real number C to ensure that

$$\|Cf(x)\| = \sqrt{\langle Cf|Cf\rangle} = C\sqrt{\int_0^L |f(x)|^2\, dx} = 1.$$

Obviously, $C = 1/\sqrt{\int_0^L |f(x)|^2\, dx}$.

Now everything starts falling in line. In the Hilbert space the functions (string shapes) are 'vectors', the 'tones' are unit basis vectors (or can be made so by the normalization), the coefficients in the trigonometric expansions are the 'components' of the 'vectors'. We can add and subtract 'vectors' and multiply them by (real or complex) numbers, we can find scalar products of two 'vectors', and we can find the 'length' (norm) of any 'vector'. Yes, this is a rather unusual 'space', but it is quite similar to the standard three-dimensional space we live in (Figure 2.7).

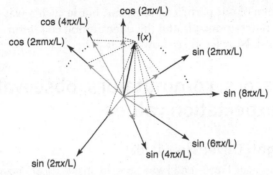

Figure 2.7 A Hilbert space of functions, which can be expanded in Fourier series

Of course, there *are* differences. For example, if you look at the expansion of $f(x)$, you see that the summation goes all the way to infinity, while a three-dimensional vector has just three components, along Ox, Oy and Oz. Well, this particular Hilbert space is *infinitely dimensional*. After dealing with 10^{23}– or so – dimensional configuration and phase spaces in Chapter 1, we should not be too shocked. There are Hilbert spaces with an even greater number of dimensions[6] – or with as few as two.

Now let us return to our state vector $\Psi(x)$. We will also denote it by $|\Psi\rangle$, to stress that we treat it as a vector in Hilbert space, and not just a function of coordinate x. Finally, we assume that it is normalized – this simplifies things:

$$\|\Psi(x)\| = \langle\Psi|\Psi\rangle^{\frac{1}{2}} = \sqrt{\int_0^L |\Psi(x)|^2\, dx} = 1.$$

We went to all the trouble of introducing a Hilbert space in order to make possible a *quantum* description of a particle in one dimension, which would be similar enough to its classical description by a vector in the phase space. That is, we demand that (i) the knowledge of the state vector $|\Psi\rangle$ at some moment of time would allow full determination of what it will be at any later time, and (ii) that $|\Psi\rangle$ would determine the position and

---

6   Yes, there exist infinities, which are larger than other infinities. This is a fascinating subject, to which Hilbert contributed as well – but unfortunately, not for this book.

momentum of our particle at any time, but in such a way that the uncertainty principle and the quantization conditions would be satisfied. Now let us see how this is done.

Operators, commutators, observables and expectation values

Spotlight: Oliver Heaviside

Oliver Heaviside (1850–1925) was an English electrical engineer, physicist and mathematician. He did not receive a formal education, but his role in the development of modern science and technology is hard to overestimate. Heaviside was a pioneer in modern communications; he co-developed vector calculus and the current form of Maxwell's equations, and gave names to a number of important parameters of electrical circuits. Heaviside obtained solutions to many key problems of electromagnetism using the technique he invented, which was based on the use of differentiation operators. When criticized for the lack of mathematical rigour, he retorted, 'Shall I refuse my dinner because I do not fully understand the process of digestion?'

Now his method is rigorously established and is known as operational calculus.

So far we have dealt with more or less conventional mathematical objects: numbers; functions; vectors; integrals – sometimes calling them unusual names (calling a function a vector, or an integral a scalar product). Now we will make a break from this routine. In order to describe quantum behaviour, we must use *operators*.

As the name suggests, an operator is a mathematical object, which describes an operation, an action of something on something.

For example, you can introduce the **operator of multiplication by the number x**, acting on the functions of x. We will denote it by a hat on top of x (a usual convention), so that

$$\hat{x} f(x) = x f(x).$$

Yes, this is just giving a fancy name to a product, but this is the simplest case. We can introduce more interesting operators, like the **differentiation operator**, which produces the derivative of a function:

$$\hat{D}f(x) = \frac{d}{dx}f(x)$$

Like conventional functions or numbers, operators can be added, subtracted or multiplied by a number. One can also multiply operators: a product of two operators, say, $\hat{A}\hat{B}$, means that in the expression $\hat{A}\hat{B}f(x)$ first we apply the operator $\hat{B}$ to $f(x)$, and then apply the operator $\hat{A}$ to the result.

Here we encounter the most evident difference between operators and usual numbers or functions: operators do not necessarily *commute*, that is, their product may depend on the order (while $2 \cdot 3 = 3 \cdot 2$ no matter what). This is not surprising; operators represent actions, and the result of two actions (e.g. turning the wheel and pushing the accelerator) may well depend on their order.

The operators $\hat{x}$ and $\hat{D}$ do not commute, because

$$\hat{x}\hat{D}f(x) = x\frac{d}{dx}f(x), \text{ while } \hat{D}\hat{x}f(x) = \frac{d}{dx}\big(xf(x)\big) = f(x) + x\frac{d}{dx}f(x).$$

In order to tell *how much* the two operators do not commute, we use their *commutator*, that is, the difference in their product due to change in the order:

$$[\hat{A}, \hat{B}] = \hat{A}\hat{B} - \hat{B}\hat{A}.$$

A commutator is denoted by square brackets.

For $\hat{x}$ and $\hat{D}$, obviously, $\hat{D}\hat{x}f(x) - \hat{x}\hat{D}f(x) = 1 \times f(x)$. This holds for any function $f(x)$, and therefore we can say that the commutator of operators $\hat{D}$ and $\hat{x}$ is

$$[\hat{D}, \hat{x}] = 1.$$

Now it is time to say what all this has to do with physics. In quantum mechanics all physical quantities – like position, velocity, electric charge and many others – are represented by operators, which *act* on the state vector of the system. This action generally changes the state vector:

$$\hat{A}|\psi\rangle = |\psi'\rangle.$$

This, by the way, is a major philosophical departure from classical physics. There the state of the system was essentially *the same*, as the set of its physical quantities. For example, the state of a classical particle is a vector in phase space with components given by the particle's position and its velocity (or momentum). In quantum mechanics, the state of a system is distinct from its physical quantities, which can be observed. (These quantities, represented by operators, are called *observables*.)

Now a natural question arises: if an operator represents a position or velocity, how can it be measured? One does not measure or observe operators. The results of our measurements or observations are, in the end, common real numbers; centimetres, degrees, kilograms. What can be observed and measured in quantum mechanics is called an *expectation value*.[7]

The expectation value of an operator depends on the state of the system. It is given by a scalar product between two vectors in the Hilbert space: the state vector *before* and *after* the operator acted upon it:

$$\langle \hat{A} \rangle = \langle \psi | \psi' \rangle = \langle \psi | \hat{A} | \psi \rangle.$$

The result is a *real* number,[8] and this is what one can actually measure or observe. ('Observation' or 'measurement' does not imply that a person is actually doing this. A trace on a light-sensitive film is as good an observation as any.)

The operators of position and momentum of a particle in one dimension are $\hat{x}$ and $\hat{p}_x = \frac{\hbar}{i} \frac{d}{dx}$. Do not be surprised: position is position, and we should expect that the Planck constant and the imaginary unit would pop up somewhere. From our exercise with operators $\hat{x}$ and $\hat{D}$ (and from the fact that $i^2 = -1$) we see that these operators do not commute:

$$\left[\hat{x}, \hat{p}_x \right] = \left[\hat{x}, \frac{\hbar}{i} \hat{D} \right] = -\frac{\hbar}{i} = i\hbar.$$

---

7 There will be yet another twist – but we will deal with that later on.

8 This is true only for certain operators and this is how we figure out whether this or that operator can correspond to a physical quantity.

Using the integral as the scalar product in our Hilbert space, we see that their expectation values are given by:

$$\langle \hat{x} \rangle = \langle \Psi | \hat{x} | \Psi \rangle = \int_0^L x |\Psi(x)|^2 \, dx \text{ and}$$

$$\langle \hat{p}_x \rangle = \langle \Psi | \hat{p}_x | \Psi \rangle = \int_0^L \Psi^*(x) \frac{\hbar}{i} \frac{d\Psi(x)}{dx} dx.$$

Key idea: Eigenstates and eigenvalues

$$\hat{A}|a\rangle = \lambda_a |a\rangle$$

$$|a\rangle$$

Generally an operator Â, acting on a state vector in a Hilbert space, changes its 'direction'.
The *eigenstates* ('own states') of this operator do not change their 'direction', but only 'length': they are simply multiplied by a real number, the corresponding *eigenvalue* ('own value') of the operator.

Eigenstates, eigenvalues, intrinsic randomness of Nature and Born's rule

We have introduced the term 'expectation value'. The explanation is that if you measure some physical quantity of a quantum system in the same quantum state several times, usually you will not get the same result! But if you repeat your measurements many times, the average of these measurements will coincide with the expectation value. Actually, the coincidence will be extremely good – quantum mechanics is arguably the theory best supported by evidence in all of science.

Spotlight: Max Born

Max Born (1882–1970) was a German physicist, Nobel Prize winner (1954) and one of the founders of quantum mechanics. His greatest contribution – Born's rule – was contained in a short footnote in his article dedicated to a particular (but now quite uninteresting) topic.

The first natural question is, if we do not measure the expectation value each time, then what do we measure – what will be the readout of our 'somethingometer'? It depends on what observable we measure and what state our system is in. Each operator of an observable has what is called a set of *eigenstates* (from the German 'own states'). They are so-called because the operator acting on them does not change their 'direction' in the Hilbert space; it just changes their 'length', by multiplying it by a real number:

$$\hat{A}|a\rangle = \lambda_a|a\rangle.$$

This real number is called the *eigenvalue* ('own value') of operator $\hat{A}$, which belongs to the eigenstate $|a\rangle$.

The measurement of any observable $\hat{A}$ always yields one of its eigenvalues, no matter what quantum state the system is in.

Do not read this as if the result of measurement did not depend on the state of the system. It does, but in a very special way.

The probability of measuring a particular eigenvalue λ_α of an observable $\hat{A}$ in a system in state $|\psi\rangle$ is given by the square modulus of the scalar product between the state $|\psi\rangle$ and state $|\alpha\rangle$:

$$P_{\lambda\alpha}(\psi) = |\langle a|\psi\rangle|^2.$$

This is known as *Born's rule*.

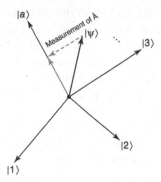

Figure 2.8 The measurement of an observable A projects the state vector on one of the eigenstates of this observable

We can make sense of this by considering the measurement as the calculation of a component of some vector along a certain direction. In conventional space, if we want to find the component of a vector R along, say, the axis Ox, we project the vector on this axis. This can be done, e.g. by calculating the scalar product between R and e_x, the unit vector along the axis. In a similar vein, we can think of the result of measurement of some observable $\hat{A}$ in state $|\psi\rangle$ as finding the projection of $|\psi\rangle$ on one of the eigenstates of the operator $\hat{A}$ (Figure 2.8). The square modulus of this projection gives the probability of observing a particular eigenvalue, while the projection of state $|\psi'\rangle = \hat{A}|\psi\rangle$, which arose because of the action of the operator $\hat{A}$ on $|\psi\rangle$), on the initial state $|\psi\rangle$ gives the average result of many repeated measurements, i.e. the expectation value $\langle\psi|\hat{A}|\psi\rangle$.

So, if a system *is* in an eigenstate (e.g. $|a\rangle$), then the measurement will give the same result, λ_α, with probability $P_{\lambda_a}(a) = |\langle a|a\rangle|^2 = 1$, that is, always. The average, of course, will be also λ_a, and it coincides with the expectation value, $\langle\hat{A}\rangle = \langle a|\hat{A}|a\rangle = \lambda_\alpha\langle a|a\rangle = \lambda_\alpha$. If this is not the case, then repeated measurements will produce at random *all* eigenvalues of $\hat{A}$ but with such probabilities that *on average* we will obtain the expectation value, $\langle\psi|\hat{A}|\psi\rangle$.

We have met with a similar situation in Chapter 1, when the ideal gas was described by a probability distribution function and the positions and velocities of particles could only be known on average. There lies the reason for such 'vagueness': there were too many particles. As we have stated there, given all the positions and velocities of all the particles in the Universe at a given moment of time, one could – in principle – exactly calculate all their future positions and velocities, and thus predict the future of the world (as well as discover all of its past). A creature capable of such an analysis was invented by Pierre-Simon, Marquis de Laplace (1749–1827), a great French astronomer, physicist and mathematician, in order to better demonstrate the determinism of classical mechanics. It would, of course, require truly supernatural powers to achieve this, and is therefore called the Laplace's demon.

Spotlight: Epicurus

Epicurus (341–270 BCE) was a Greek atheist philosopher. He further developed the atomistic philosophy of Democritus and suggested that from time to time atoms in motion undergo spontaneous *declinations* – what can be considered as the predecessors of random quantum leaps!

The similarity does not go all the way. Quantum mechanics is different. The randomness of outcomes is *fundamental*. It is built into the very foundations of Nature and has nothing to do with our human shortcomings. No Laplacian demon will be able to get rid of it. It may well seem strange – but this is the way it is. Moreover, even at our macroscopic level we are too well acquainted with chance to think it wholly unnatural. Epicurus explicitly considered the concept of Nature's intrinsic randomness as early as the 3rd century BCE. And this randomness has everything to do with quantization and the uncertainty principle.

Quantization and Heisenberg uncertainty relations

In order to properly describe Nature, we made some serious changes to its classical description in Chapter 1. We have separated the state of system from physical quantities (observables), which can be observed in this system. We have described the state by a vector in a very unusual (Hilbert) space, and observables by operators, acting on this vector.

We have stated that Nature is intrinsically random. We have related measurement of physical quantities with projections of the state vector on some other vectors (the eigenstates of a given observable) in Hilbert space. How will this help us with quantization and uncertainty relation?

Earlier on in this chapter we have stated that for a particle in one dimension, position and momentum are represented by operators $\hat{x}$ and $\hat{p}_x = \frac{\hbar}{i}\frac{d}{dx}$, and that the commutator of these operators is

$$\left[\hat{x}, \hat{p}_x\right] = i\hbar.$$

They say that $\hat{x}$ and $\hat{p}_x$ *commute to $i\hbar$*. Recall that in classical mechanics position and momentum are conjugate variables. The 'recipe' for producing a quantum description of a system is precisely this:

Take the conjugate variables and replace them by operators, which commute to $i\hbar$.

That is it. We will start to discuss in the next chapter what to do afterwards. But we can deal with the uncertainty principle right now.

Note that to know the average is not enough. Suppose we measure the position of a particle, and the result is with equal probability either +100 m or –100 m. After taking a million measurements and averaging, we will, of course, obtain the average position zero, but the particle will be never even close to this point. It is necessary to know the *uncertainty* of the result. The simplest way of finding it is to calculate the *standard deviation*, that is, the average mean root square of the difference between the particular measurement and the average. In application to an observable $\hat{A}$, this is given by

$\Delta A = \sqrt{\langle(\hat{A} - \langle\hat{A}\rangle)^2\rangle}$. From a naïve application of the uncertainty principle, one could think that it will be always finite and somehow determined by the Planck constant – something like drawing phase trajectories of finite length. We saw that this will not do, and our Hilbert space description is subtler than that. We have just seen that if the system is in an eigenstate of $\hat{A}$, then the result of measurement is always the same, and $\Delta A = 0$.

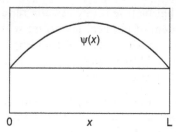

Figure 2.9 The wave function of a particle in a one-dimensional box (in the lowest-energy state)

Now consider the state $|\psi\rangle$ of our particle in a 'box of length L', given by $\Psi(x) = \sqrt{\frac{2}{L}} \sin\frac{\pi}{L}x$ (Figure 2.9). The norm $\|\Psi(x)\| = \sqrt{\langle \Psi | \Psi \rangle} = \sqrt{\int_0^L |\Psi(x)|^2\, dx} = 1$, so this is an acceptable state vector. Let us ask two questions: (i) What will be the expectation value of position and momentum of such a particle, i.e. the averaged measurements of these quantities? (ii) What is the uncertainty of these measurements?

These are both easy to calculate:

$$\langle \hat{x} \rangle = \int_0^L x\,|\Psi(x)|^2\, dx = \frac{L}{2};$$

$$\Delta x = \sqrt{\int_0^L \left(x - \frac{L}{2}\right)^2 |\Psi(x)|^2\, dx} = \frac{L}{2\pi}\sqrt{\frac{\pi^2 - 6}{3}} \approx 0.36\frac{L}{2};$$

$$\langle \hat{p}_x \rangle = \int_0^L \Psi^*(x)\frac{\hbar}{i}\frac{d}{dx}\Psi(x)\, dx = 0;$$

$$\Delta p_x = \sqrt{\int_0^L \Psi^*(x)\left(\frac{\hbar}{i}\right)^2 \frac{d^2}{dx^2}\Psi(x)} = \frac{\pi\hbar}{L};$$

We see that $|\Psi\rangle$ is not an eigenstate of position or momentum. The measurements of either produce uncertainties, which depend on the length L. What *is* interesting though is that the product of these uncertainties only depends on the fundamental Planck constant:

$$\Delta x \ \Delta p_x \approx \frac{0.36\pi}{2}\hbar \approx 0.57\hbar.$$

This is a particular case of a very general and rigorous statement, the *Heisenberg uncertainty relation*:

For any two conjugate observables (in other words, commuting to $i\hbar$) the product of their uncertainties in any quantum state cannot be less than $\hbar/2$.

$$\text{For example } \Delta x \ \Delta p_x \geq \frac{\hbar}{2}$$

In this way we have managed to introduce quantum uncertainty and quantization without appealing to phase trajectories. It was hard going, but from now on the road will become easier.

Fact-check

1 The energy of a harmonic oscillator can change by

  a $h\nu$

  b $h\omega$

  c $\hbar\nu$

  d $\hbar\omega$

2 Action is

  a for example, the area enclosed by a phase trajectory

  b the effect produced by a quantum system on its surroundings

  c the effect produced on a quantum system by its surroundings

  d a physical quantity introduced by Max Planck

3 The Planck constant is called the quantum of action, because

  a action can only change by a multiple of h

  b action is a multiple of h

  c action of a harmonic oscillator can only change by a multiple of h

  d this is the smallest value action can have

4 An adiabatic invariant of a harmonic oscillator is

    a   its frequency

    b   ratio of its energy to frequency

    c   its energy

    d   its action

5 Bohr–Sommerfeld quantization condition applies to

    a   only one-dimensional harmonic oscillator

    b   only one-dimensional mechanical systems

    c   any mechanical system with closed phase trajectories

    d   only mechanical systems

6 Can a physical system with action much larger than h behave as a quantum object?

    a   no

    b   yes

    c   no, such a system is not described by quantum mechanics

    d   only if it is a harmonic oscillator

7 Hilbert space is

    a   an abstract space, where the state vector of a quantum system 'lives'

    b   an isolated space where a quantum system must be placed in order to observe its properties

    c   the usual space as it is called by physicists

    d   the part of the Universe where quantum mechanics applies

8 In quantum mechanics, physical quantities are represented by

    a   vectors

    b   operators

    c   commutators

    d   expectation values

9 Conjugate variables in quantum mechanics are represented by

    a   operators with a commutator equal to zero

    b   operators with a commutator equal to $i\hbar$

    c   eigenvectors

    d   operators with a commutator equal to $\left(n+\frac{1}{2}\right)\hbar$

10 Which of the below expressions is correct?

a $\Delta x\ \Delta p_x \geq \dfrac{\hbar}{2}$

b $\Delta x\ \Delta p_x \geq \dfrac{h}{2}$

c $\Delta x\ \Delta p_z \geq \dfrac{\hbar}{2}$

d $\Delta E\ \Delta \omega \geq \dfrac{\hbar}{2}$

Dig deeper

Further reading I. Basics of quantum mechanics.

R. Feynman, R. Leighton and M. Sands, *Feynman Lectures on Physics*, Vols I, II and III. Basic Books, 2010 (there are earlier editions). See Vol. I (Chapters 37, 41 and 42) and Vol. III (start with Chapters 1, 3, 5 and 8).

E. H. Wichmann, *Quantum Physics* (Berkeley Physics Course, Vol. 4). McGraw-Hill College, 1971. This is a university textbook, but it is probably the most accessible of the lot.

Further reading II. About the ideas and scientists.

J. Baggott, *The Quantum Story: A History in 40 Moments*. Oxford University Press, 2011. A modern history. Well worth reading.

G. Gamow, *Thirty Years that Shook Physics: The Story of Quantum Theory*. Dover Publications, 1985. A history written by an eyewitness and one of the leading actors in the field.

3

Equations of quantum mechanics

Long ago the great mathematician and physicist Laplace presented his definitive work on celestial mechanics to Napoleon. Accepting the book, Napoleon – who was both curious and knowledgeable about the latest scientific developments – asked whether it was true that in such an important book about the world system there was no mention of the world's creator. Laplace famously replied: 'Sire, I had no need of that hypothesis.' In a less famous sequel to this exchange, another great mathematician and physicist, Lagrange, commented: 'Ah, but it is such a fine hypothesis; it explains many things,' to which Laplace allegedly answered: 'This hypothesis, Sire, indeed explains everything, but permits to predict nothing. As Your Majesty's scholar, I must provide you with works permitting predictions.'

Science is about explanations, which do permit predictions (and the success of an explanation is measured by its predictive power). In physics, to predict the future is to solve the *equations of motion*, which link the future state of a physical system to its present (and sometimes its history as well).

When introducing the Hilbert space, state vector and operators, we claimed that this will allow us to keep some essential features of classical mechanics – such as the description of a quantum system's evolution in terms of a vector – in some exotic, but comprehensible, space so that once we know the state of the system, we can predict it at all future times.

The price we paid is pretty high. For a single particle in one dimension instead of a two-dimensional phase space we now have to deal with an infinite-dimensional Hilbert space! Therefore let us check that we at least obtained what we have bargained for.

A part of the price was the 'split' between the state of the system, now represented by the vector $|\Psi\rangle$, and the observables; in this case the position and momentum operators, $\hat{x}$ and $\hat{p}_x$. This means that when considering the evolution of a quantum system with time, we could either look at the change in $|\Psi\rangle$, or in $\hat{x}$ and $\hat{p}_x$. Both approaches are equally valid and both are being used, depending on what is more convenient. The first is due to Schrödinger, the second due to Heisenberg, but it was Dirac who realized that they are actually the same and that they are closely related to classical mechanics.

The similarity is easier to see if we start with the *Heisenberg representation*.

Spotlight: Paul Dirac

Paul Adrien Maurice Dirac (1902–1984) was an English physicist and Nobel Prize winner (1933).

Among other important results, he predicted the existence of antimatter and precisely described its properties. The Dirac equation, which he discovered, describes both electrons and positrons in agreement with Einstein's special relativity. Dirac reinvented the so-called Dirac's delta function $\delta(x)$ (an infinitely sharp pulse), which was before used by Heaviside.

Dirac was famous for his precision of speech. Once during his talk he did not react to somebody's 'I do not understand the equation on the top-right-hand corner of the blackboard'. After being prompted by the moderator to answer the question, Dirac said: 'That was not a question, it was a comment.'

Poisson brackets, Heisenberg equations of motion and the Hamiltonian

We recall Hamilton's equations for a classical particle in one dimension:

$$\frac{dx}{dt} = \frac{\partial}{\partial p_x} H(x, p_x); \frac{dp_x}{dt} = -\frac{\partial}{\partial x} H(x, p_x).$$

They can be rewritten using the *Poisson bracket*. In classical mechanics, the Poisson bracket $\{f, g\}$ that combines the two functions of position and conjugate momentum is called the following combination of derivatives:

$$\{f, g\} = \frac{\partial f}{\partial x} \frac{\partial g}{\partial p_x} - \frac{\partial g}{\partial x} \frac{\partial f}{\partial p_x}.$$

If there is more than one pair of conjugate variables (e.g. for N particles in three dimensions), one only needs to add up the contributions from all these pairs:

$$\{f, g\} = \sum_{n=1}^{N} \sum_{\alpha = x, y, z} \left(\frac{\partial f}{\partial \alpha_n} \frac{\partial g}{\partial p_{\alpha n}} - \frac{\partial g}{\partial \alpha_n} \frac{\partial f}{\partial p_{\alpha n}} \right).$$

Poisson brackets have many important uses, but what we need here is to rewrite Hamilton's equations as

$$\frac{dx}{dt} = \{x, H\}; \frac{dp_x}{dt} = \{p, H\}.$$

We use the fact that the variables x and p_x are independent of each other. In other words, the partial derivatives $\frac{\partial x}{\partial p_x} = 0$ and $\frac{\partial p_x}{\partial x} = 0$. Remember that when we take a partial derivative with respect to one independent variable, we keep all others constant. Then indeed

$$\frac{dx}{dt} = \frac{\partial x}{\partial x}\frac{\partial H}{\partial p_x} - 0 = \frac{\partial H}{\partial p_x}; \frac{dp_x}{dt} = 0 - \frac{\partial H}{\partial x}\frac{\partial p_x}{\partial p_x} = -\frac{\partial H}{\partial x}.$$

Moreover, for any function f of x, p_x (e.g. a probability distribution function) its time derivative will be, according to the chain rule and the Hamilton's equations, a Poisson bracket of this function with the Hamilton function:

$$\frac{df(x,p_x)}{dt} = \frac{\partial f}{\partial x}\frac{dx}{dt} + \frac{\partial f}{\partial p_x}\frac{dp_x}{dt} = \frac{\partial f}{\partial x}\frac{\partial H}{\partial p_x} + \frac{\partial f}{\partial p_x}\left(-\frac{\partial H}{\partial x}\right) = \{f, H\}.$$

If there are many pairs of conjugate variables, the formula $\frac{df}{dt} = \{f, H\}$ remains the same. Only the Poisson bracket will now be the sum of contributions from all these pairs.

The important *physical* insight from this equation is that the Hamilton function – that is, energy – determines the time evolution of any physical system. Time and energy turn out to be related in classical mechanics, almost like (but not quite) the position and momentum.

What has this to do with quantum mechanics? A while ago Werner Heisenberg found out that *quantum equations of motion for operators (Heisenberg equations)* have the following form:[1]

$$i\hbar\frac{d\hat{A}}{dt} = [\hat{A}, \hat{H}].$$

---

1   Of course, Heisenberg did not *derive* these equations – there was nothing to derive them from. He *postulated* them, just as Newton postulated his laws of motion, based on some theoretical insights and experimental data, and with the eventual justification that they produce correct (that is, agreeing with the experiment) answers. On the other hand, the Hamilton's equations *are* derived from Newtonian laws – or, if you wish, the other way around: they are equivalent.

In particular, $i\hbar\dfrac{d\hat{x}}{dt} = [\hat{x}, \hat{H}]$; $i\hbar\dfrac{d\hat{p}_x}{dt} = [\hat{p}_x, \hat{H}]$. Here the brackets denote the commutator of two operators.

Key idea: Rotations

$$r \rightarrow r' = \hat{O}\,r \qquad \begin{pmatrix} x' \\ y' \end{pmatrix} = \begin{pmatrix} \cos\alpha & -\sin\alpha \\ \sin\alpha & \cos\alpha \end{pmatrix}\begin{pmatrix} x \\ y \end{pmatrix}$$

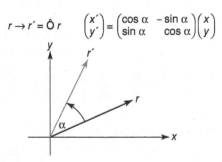

A vector **r** can be rotated to become vector **r'**. This operation can be written as the action of the *rotation operator* $\hat{O}$ on vector **r:**

$$r' = \hat{O}\,r.$$

It transforms the old coordinates (e.g. *(x,y,z)*) into the new ones *(x',y',z')*.

This is a particularly simple operation in a two-dimensional space; a planar rotation by an angle α. Then the vector **r** can be represented as a column of two numbers, *x* and *y*, and the operator $\hat{O}$ as a two-by-two matrix.

The Hamiltonian operator $\hat{H}$ is the quantum operator, which corresponds to the classical Hamilton function, i.e. the energy of the system. For example, for a harmonic oscillator it is simply

$$\hat{H} = \dfrac{\hat{p}_x^2}{2m} + \dfrac{m\omega_0^2\hat{x}^2}{2}.$$

One of the great insights of Paul Dirac was that these *Heisenberg equations of motion* are directly analogous to classical equations of motion written using Poisson brackets. One only needs to replace the classical functions and variables with quantum operators, and instead of a classical Poisson bracket put in the commutator divided by $i\hbar$:[2]

$$\{f,g\} \rightarrow \frac{1}{i\hbar}[\hat{f},\hat{g}],$$

And voilà! You have a quantum equation of motion. This is a much more consistent scheme of quantization than the one of Bohr–Sommerfeld.

Once we know the solutions to the Heisenberg equations, the expectation value for an observable $\hat{A}(t)$ at any moment of time is immediately found:

$$A(t) = \langle \hat{A}(t) \rangle = \langle \Psi \mid \hat{A}(t) \mid \Psi \rangle.$$

Here $|\Psi\rangle$ is the state vector of our system. Note that it does not change in time. This is because in the *Heisenberg representation* we consider the time dependence of the operators themselves. The state vector is static. We can do this due to the aforementioned 'split' between the state vector and the operators of physical quantities.

Matrices and matrix elements of operators

Key idea: Matrices

$$\hat{M} = \begin{pmatrix} m_{11} & m_{12} & m_{13} & \cdots \\ m_{21} & m_{22} & m_{23} & \cdots \\ & & \cdots \end{pmatrix}$$

$$\hat{M}a = \begin{pmatrix} m_{11} & m_{12} & m_{13} & \cdots \\ m_{21} & m_{22} & m_{23} & \cdots \\ & & \cdots \end{pmatrix} \begin{pmatrix} a_1 \\ a_2 \\ \cdots \end{pmatrix} = \begin{pmatrix} m_{11}a_1 + m_{12}a_2 + \cdots \\ m_{21}a_1 + m_{22}a_2 + \cdots \\ \cdots \end{pmatrix}$$

$$\hat{M}\,\hat{Q} = \begin{pmatrix} m_{11} & m_{12} & \cdots \\ m_{21} & m_{22} & \cdots \\ & \cdots \end{pmatrix} \begin{pmatrix} q_{11} & q_{12} & \cdots \\ q_{21} & q_{22} & \cdots \\ & \cdots \end{pmatrix} = \cdots$$

2   If you want to go back to classical mechanics, replace the operators with functions and put $\hbar \rightarrow 0$. The commutator in the numerator will then go to zero, but so will the denominator, and their ratio will revert to the classical Poisson bracket.

A matrix product is a basic operation of linear algebra. Two square matrices can be multiplied using the 'row-by-column' rule and produce another square matrix.

This operation is similar to the usual multiplication of numbers (e.g. it is distributive and associative), but it is not commutative:

$$\hat{A}\hat{B} \neq \hat{B}\hat{A}$$

Using the same rule, we can multiply a vector column by a square matrix *from the left*, producing another vector column.

You can check that one can multiply a vector row by a square matrix *from the right* and obtain another vector row.

In quantum mechanics vector columns are denoted by 'kets', $|\psi\rangle$, and vector rows by 'bras', $\langle\psi|$. Then an operator, represented by a square matrix, can act on a 'ket' from the left, and on a 'bra' from the right:

$$\hat{A}|\psi\rangle = |\psi'\rangle,$$
$$\langle\psi|\hat{A} = \langle\psi'|.$$

It is easy to say 'once we know the solution', but how are we to solve equations, which involve operators, and then to calculate quantities like $\langle\Psi|\hat{A}(t)|\Psi\rangle$? Here we are yet again helped by an analogy with geometry, vectors and spaces (Figure 3.1).

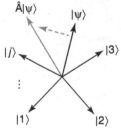

Figure 3.1 Action of an operator in a Hilbert space

An operator $\hat{A}$ acts on a vector $|\Psi\rangle$ in a Hilbert space: $\hat{A}|\Psi\rangle = |\Psi'\rangle$.

We can expand both state vectors, $|\Psi\rangle$ and $|\Psi'\rangle$:

$$|\Psi\rangle = \sum \psi_j \,|j\rangle; |\Psi'\rangle = \sum \psi_j' \,|j\rangle.$$

Here $|j\rangle$ are basis states of the Hilbert space (like sine and cosine functions in Chapter 2), which play the same role as unit vectors

e_x, e_y, e_z in our conventional three-dimensional space. Depending on what system we are dealing with, there can be finite or infinite number of such basis states, but this will make no difference.

Recall that the scalar product of two different basis vectors is zero: $\langle j|k \rangle = 0$, $j \neq k$. Of course, the scalar product of a basis vector with itself is the square of its length, that is, one: $\langle j|j \rangle = \|j\|^2 = 1$.

The notation we use for the scalar products of vectors in a Hilbert space, $\langle b|a \rangle$, is another contribution due to Dirac. It can be interpreted as a product of a vector $|a\rangle$ ('ket-vector', or simply 'ket') and a 'conjugate' vector $\langle b|$ ('bra-vector', or 'bra'). Together they form a bracket, so the terms are self-explanatory. A bra $\langle b|$ is conjugate to the ket $|b\rangle$: if $|b\rangle$ is described by some complex coefficients, then $\langle b|$ is described by complex conjugates of these coefficients. Operators act on a ket vector from the left: $\hat{A}|a\rangle$, but on the bra vectors from the right: $\langle a|\hat{A}$.

A state vector is a ket: $|\Psi\rangle = \sum \psi_j\, |\,j\rangle$; the corresponding bra is $\langle \Psi | = \sum \psi_j^* \langle j\,|$. They contain the same information about the system.

The coefficients ψ_j, ψ'_j (components of the state vector before and after it was acted upon by the operator $\hat{A}$) are just some complex numbers. Therefore all we need to know about the operator $\hat{A}$ is how it changes these into the others.

In conventional space the action of operators is well known. (They are not necessarily called operators, but this is a question of terminology.) For example, a rotation of a vector in xy-plane by the angle θ is given by an operator $\hat{O}(\theta)$. In a Hilbert space of quantum mechanics we follow the same approach. An operator can be written as a *matrix,* a square table of numbers:

$$\hat{A} = \begin{pmatrix} A_{11} & A_{12} & \cdots \\ A_{21} & A_{22} & \cdots \\ \cdots & \cdots & \cdots \end{pmatrix} = \left(A_{ij} \right),$$

It has as many rows (columns), as there are basis vectors in our (usual or unusual) space. The action of our operator on the state vector is then given by the *matrix product*:

$$| \Psi' \rangle = \sum_j \psi_j | j \rangle = \hat{A} | \Psi \rangle = \sum_j \left(\sum_k A_{jk} \psi_k \right) | j \rangle,$$

or, keeping just the expression for the state vector components,

$$\psi'_j = \sum_k A_{jk} \psi_k.$$

The rule for multiplying two matrices, e.g. $\hat{A}$ and $\hat{B}$ is the same: the (jk)th matrix element of the product $\hat{C} = \hat{A}\,\hat{B}$ is given by the same 'row-by-column' rule:

$$C_{ij} = \sum_k A_{ik} B_{kj}.$$

Note that for so defined matrix product $\hat{A}\,\hat{B} \neq \hat{B}\,\hat{A}$! This is, of course, exactly what we would want for operators.[3] Note also that nothing depends on the index k on the right hand side, since it is summed over. We could denote it by any other letter. Such indices, over which summation is taken, are called *dummy* indices.

Since the components of state vectors and the matrix elements of operators are conventional (though complex) numbers, now we can deal with the Heisenberg equations. We only need to write all operators as matrices and deal with them element by element. The commutator $[\hat{A}, \hat{H}] = \hat{A}\hat{H} - \hat{H}\hat{A}$ is now a matrix, with the matrix element

$$[\hat{A}, \hat{H}]_{ij} = \sum_k (A_{ik} H_{kj} - H_{ik} A_{kj}).$$

The Heisenberg equations thus become

$$i\hbar \frac{dA_{ij}}{dt} = \sum_k (A_{ik} H_{kj} - H_{ik} A_{kj}).$$

---

3   Mathematicians figured it out in the 19th century. Heisenberg did not study this particular branch of mathematics (linear algebra) and had to rediscover it all by himself. Now it is a prerequisite for studying quantum mechanics.

This is not a single equation, but a set of equations for all pairs (i, j). This may be even an infinite set, but such equations can be solved either exactly, or more often approximately, or even more often, numerically. The point is, such a solution (a set of numbers, the time-dependent matrix elements $A_{ij}(t)$) can be obtained.

Now we can calculate the expectation value of the observable $\hat{A}$,
$$\langle \hat{A}(t) \rangle = \langle \Psi | \hat{A}(t) | \Psi \rangle .$$

We know that the vector $| \Psi'(t) \rangle = \hat{A}(t) | \Psi \rangle = \sum_j \left(\sum_k A_{jk}(t)\psi_k \right) | j \rangle$. It remains to find its scalar product with the vector $| \Psi \rangle = \sum_i \Psi_i | i \rangle$:

$$\langle \psi | \psi' \rangle = \sum_i \psi_i^* \langle i | \sum_j \left(\sum_k A_{jk}(t)\psi_k \right) | j \rangle = \sum_i \sum_j \sum_k \psi_i^* \psi_k A_{jk}(t)\langle i | k \rangle.$$

Since $\langle i|k \rangle = 1$ if $i = k$ and zero otherwise, one summation can be carried out immediately, and we obtain:

$$\langle \hat{A}(t) \rangle = \sum_i \sum_j \psi_i^* \psi_j A_{ji}(t).$$

This formula contains complex numbers. But we cannot directly measure complex numbers. Therefore the matrix elements $Aji(t)$ must be such, that for *any* set of complex numbers ψ_j, forming the components of some state vector, the expectation value was *real*. We have already mentioned that not every operator can be an operator of a physical quantity – well, this is the main requirement, which must be satisfied. Fortunately, it is pretty easy to satisfy and we don't need to dwell on it.

The approach, based on matrix representations of operators and developed by Heisenberg, was for obvious reasons called *matrix mechanics*, but the term is now obsolete because this is just one particular way of writing down equations of quantum mechanics.

Schrödinger equation and wave function

Spotlight: Schrödinger's cat

Erwin Schrödinger (1887–1961) was an Austrian physicist, Nobel Prize winner (1933) and one of the creators of quantum mechanics.

His public fame is more due to the *Schrödinger's cat* – a paradoxically alive-and-dead animal from a thought-experiment *(Gedankenexperiment)* he invented to underline one of the key 'paradoxes' of quantum mechanics.

We have dealt with the evolution of a quantum system in the Heisenberg representation, i.e. by tracing the time dependence of operators, which describe the measurable physical quantities of the system, and letting its quantum state vector stay put. But we can instead let it go and fix the operators. The main thing is that the expectation values – things, which can be actually observed and measured – should stay the same. When it is the state vector that changes with time while the operators are time-independent, the approach is called the *Schrödinger representation*.

Let us calculate the time derivative of the expectation value written as $\langle \Psi | \hat{A} | \Psi \rangle$. Using the standard rule for the derivative of a product, we find:

$$\frac{d\langle \hat{A} \rangle}{dt} = \left(\frac{d}{dt} \langle \Psi | \right) \hat{A} | \Psi \rangle + \langle \Psi | \left(\frac{d}{dt} \hat{A} \right) | \Psi \rangle + \langle \Psi | \hat{A} \left(\frac{d}{dt} | \Psi \rangle \right).$$

In Heisenberg representation $|\Psi\rangle$ and $\langle\Psi|$ are time independent, their time derivatives are zero and the time derivative of the operator is given by the Heisenberg equations, that is,

$$\frac{d\langle \hat{A} \rangle}{dt} = \langle \Psi | \left(\frac{1}{i\hbar} \hat{A}\hat{H} - \frac{1}{i\hbar} \hat{H}\hat{A} \right) | \Psi \rangle = -\left(\langle \Psi | \hat{H} \frac{1}{i\hbar} \right) \hat{A} | \Psi \rangle + \langle \Psi | \hat{A} \left(\frac{1}{i\hbar} \hat{H} | \Psi \rangle \right).$$

On the other hand, if the operators are fixed and the state vector changes in time, we get

$$\frac{d\langle\hat{A}\rangle}{dt} = \left(\frac{d}{dt}\langle\Psi|\right)\hat{A}|\Psi\rangle + \langle\Psi|\hat{A}\left(\frac{d}{dt}|\Psi\rangle\right).$$

The two equations agree, if $\frac{d}{dt}|\Psi\rangle = \frac{1}{i\hbar}\hat{H}|\Psi\rangle$ and $\frac{d}{dt}\langle\Psi| = -\frac{1}{i\hbar}\langle\Psi|\hat{H}$.

These two relations are actually equivalent (they are conjugate). Therefore we have derived the famous *Schrödinger equation*:

$$i\hbar\frac{d}{dt}|\Psi\rangle = \hat{H}|\Psi\rangle.$$

We derived it starting from the Heisenberg equations. Schrödinger himself postulated it based on some theoretical insight and experimental data, completely independently of Heisenberg. The two approaches were initially considered different and competing theories, until Dirac demonstrated that they are actually the same theory – quantum mechanics – written in two different ways. In the 'Heisenberg representation' the operators are time-dependent and state vectors static; in the 'Schrödinger representation' it is the other way around. But the expectation values of the physical quantities are, of course, the same in both representations.[4]

The first applications of quantum theory were to microscopic particles, such as electrons and atoms, and their interaction with the electromagnetic field. The Schrödinger equation happened to be more convenient for dealing with these problems and therefore was more widely used. As a result, some of the related terminology survives in modern quantum mechanics. For example, the quantum state vector is often called the *wave function* (usually if it describes the quantum state as a function of coordinates in our usual, three-dimensional space).

The relative simplicity of the Schrödinger equation comes from the fact that usually you can substitute into it the state vector $|\Psi\rangle$

---

4   Dirac showed that the difference between the Schrödinger and Heisenberg representations is like the one between different frames of references in classical mechanics, and the choice of the one or the other is purely a question of convenience. He also invented yet another representation (the *interaction representation*), but unless you want to become a professional physicist you do not need to bother with it.

directly in the form of a conventional function of position (in a one-dimensional case, $\Psi(x)$) – which is what we had in Chapter 2 *before* presenting it as an expansion over basis functions.

Consider for starters the case of a *free* particle of mass m in one dimension. The Hamiltonian of a free particle contains only a kinetic energy term since there is no potential acting on it:

$$\hat{H} = \frac{\hat{p}^2}{2m} = \frac{1}{2m}\left(\frac{\hbar}{i}\frac{d}{dx}\right)^2.$$

Now the Schrödinger equation looks like

$$i\hbar\frac{\partial}{\partial t}\Psi(x,t) = \frac{1}{2m}\left(\frac{\hbar}{i}\frac{\partial}{\partial x}\right)^2\Psi(x,t) = -\frac{\hbar^2}{2m}\frac{\partial^2}{\partial x^2}\Psi(x,t).$$

We have replaced here full (straight) derivatives with partial (curved) ones, since the state of the system depends on two variables at once: position and time. This is the simplest example of a *wave equation*, a class of equations thoroughly investigated in the 18th and 19th centuries in relation to the problems of mechanics and optics.[5] We can immediately write a solution to this equation:

$$\Psi(x,t) = Ce^{ikx}e^{-i\omega t}.$$

Using the property of an exponential function that $\frac{de^{az}}{dz} = ae^{az}$, and therefore $\frac{d^2 e^{az}}{dz^2} = a^2 e^{az}$, you can check that this is indeed a solution, if

$$\hbar\omega = \frac{(\hbar k)^2}{2m}.$$

What does this solution look like? Using Euler's formula, we see that (assuming C is a real constant)

$$\Psi(x,t) = C\{\cos(kx - \omega t) + i\sin(kx - \omega t)\}.$$

---

5 Hamilton made key contributions to both, and his works establishing similarities between optics and classical mechanics played an important role in the discovery and development of quantum mechanics.

Both the real and the imaginary parts of $\Psi(x,t)$ are *waves*, which have frequency ω and *wave vector k* and propagate with the speed $s = \dfrac{\omega}{k}$ (Figure 3.2). Naturally, Ψ is called a wave function. The *wavelength* λ (the distance between the consecutive maxima or minima) is found from the condition $\cos (k(x + \lambda)) = \cos kx$ and equals $\lambda = \dfrac{2\pi}{k}$.[6]

In an oscillator, the combination $\hbar\omega$ is equal to its energy. This holds true here as well (after all, a free particle is just an oscillator with a broken spring): the energy of a particle is

$$E = \hbar\omega.$$

We can therefore write

$$\Psi(x, t) = \Psi(x)e^{-\frac{iE}{\hbar}t}.$$

This way we separate the *stationary* part of the wave function that depends only on position, from the part that describes its time dependence. This can be done for any quantum system, *which has a definite energy E*. Substituting such a wave function in the Schrödinger equation, we obtain the *stationary* Schrödinger equation:

$$E\Psi(x) = -\frac{\hbar^2}{2m}\frac{d^2}{dx^2}\Psi(x).$$

In case of more than one dimension, the change to the Schrödinger equations is trivial. For example, if a particle travels in space – our usual, three-dimensional space – we will either be solving the equation

$$i\hbar\frac{\partial}{\partial t}\Psi(x,y,z,t) = -\frac{\hbar^2}{2m}\left(\frac{\partial^2}{\partial x^2} + \frac{\partial^2}{\partial y^2} + \frac{\partial^2}{\partial z^2}\right)\Psi(x,y,z,t),$$

or – if the system has a definite energy E and its wave function can be written as $\Psi(x,y,z)e^{-\frac{iEt}{\hbar}}$ – the equation

---

6   The particular solution $\Psi(x,t)$ with definite frequency (= energy) and wave vector is called a *plane wave*.

$$E\Psi(x,y,z) = -\frac{\hbar^2}{2m}\left(\frac{\partial^2}{\partial x^2} + \frac{\partial^2}{\partial y^2} + \frac{\partial^2}{\partial z^2}\right)\Psi(x,y,z).$$

The solutions to the stationary Schrödinger equation are called the *stationary states*.

Quantum propagation

One conceptual problem with the use of a wave function is how a wave can describe the propagation of a particle in a definite direction. Here again much can be inferred from classical physics – this time from optics.

Spotlight: Christiaan Huygens

Christiaan Huygens (1629–1695) was a Dutch scientist who, among other important discoveries, formulated an early version of the wave theory of light, which allowed him to correctly explain the processes of reflection and refraction as well as the birefringence in crystals.

Spotlight: The Poisson-Arago-Fresnel spot

A French physicist, Augustin-Jean Fresnel, further developed the wave theory of light and showed how it can be used to explain light propagation along a straight line.

Siméon Denis Poisson, another French physicist, was present at Fresnel's presentation of his results and immediately objected that – if this theory is right – there always must be a small light spot in the centre of a shadow cast by a round screen, which clearly makes no sense.

Another physicist who attended the presentation, François Arago, right after the meeting went to his laboratory to conduct the experiment – and observed the spot, predicted by Poisson. At the next meeting he reported his discovery, which is now called the Fresnel, Poisson, or Arago spot. It decided the argument about the wave nature of light.

The wave theory of light was well established in the 19th century, though it hails back to Christiaan Huygens, a Dutch contemporary and competitor of Newton. In the process of its development, the question of how a wave can propagate as a ray of light was successfully answered (with a contribution from Hamilton). Roughly speaking, if the wavelength is small enough, one can shape the waves into a beam in a definite direction.

Consider again the one-dimensional sinusoidal wave shape, for example, $\sin(kx - \omega t)$, shown in Figure 3.2. The sinusoid moves to the right with the speed $s = \omega/k$. Since on the one hand $E = \frac{(\hbar k)^2}{2m}$, and on the other (from classical mechanics), $E = \frac{p^2}{2m}$, the momentum of a free particle is proportional to its wave vector and inversely proportional to its wavelength:

$$p = \hbar k = \hbar \frac{2\pi}{\lambda} = \frac{h}{\lambda}.$$

Is this consistent with our earlier definition of quantum mechanical momentum, $\hat{p} = \frac{\hbar}{i} \frac{d}{dx}$? Let us calculate the expectation value of this operator for a free particle:

$$\langle \hat{p} \rangle = \langle \Psi | \hat{p} | \Psi \rangle = \int_{-\infty}^{\infty} \{ C^* e^{\frac{iEt}{\hbar}} e^{-\frac{ipx}{\hbar}} \frac{\hbar}{i} \frac{d}{dx} \left(C e^{\frac{-iEt}{\hbar}} e^{\frac{ipx}{\hbar}} \right) \} dx = |C|^2 \, p \int_{-\infty}^{\infty} e^{-\frac{ipx}{\hbar}} e^{\frac{ipx}{\hbar}} \, dx$$

$$= |C|^2 \, p \int_{-\infty}^{\infty} 1 \, dx.$$

How to deal with the infinitely large integral on the right-hand side? We demanded that the state vectors were normalized, that is, that $\| \Psi \|^2 = \langle \Psi | \Psi \rangle = 1$. This is why we included in our wave function the constant factor C. In order to normalize a plane wave, we must put $|C|^2 = \frac{1}{\int_{-\infty}^{\infty} 1 \, dx}$. Formally it is all right: the infinitely large integrals cancel, and we immediately find, that indeed for a plane-wave wave function

$$\langle \hat{p} \rangle = p.$$

Dealing with infinitely large or small coefficients is, of course, unpleasant, but this kind of trouble is not specific to quantum mechanics and was encountered in theory of classical waves. The infinities come from 'unnatural' character of a plane wave solution. Such a wave, strictly speaking, must permeate *all* of space at *all* times – which is, to put it mildly, impossible. In real life one must deal with what is called *wave packets* of finite extension and duration, and then the infinities do not occur – but the calculations become messier.

By the way, this is an extreme example of quantum uncertainty principle. The *momentum* of a particle in a state, which is given by a plane-wave wave function, is precisely known: $p = \hbar k$ You can check directly that $\Delta p = 0$. On the other hand, since the wave exists in all of the space at once and is the same everywhere, the *position* of the particle is indefinite, that is $\Delta x = \infty$.

A particle in a box and on a ring: energy and momentum quantization

The problems with infinitely large or small coefficients disappear if instead of unlimited space (or, in a one-dimensional case, a line, $-\infty < x < \infty$) we put our quantum particle in a box (in a one-dimensional case, on a segment, $0 \leq x \leq L$; see Figure 2.9). This means that the wave function must be zero on the sides and outside the box.

In one dimension, we will seek the answer in the form
$\Psi_n(x) = C_n \sin \dfrac{\pi n x}{L}$ on the segment $0 < x < L$, and $\Psi_n(x) = 0$, outside. Here n is an integer, and the coefficient C_n is chosen so as to normalize the wave function. Substituting this in the stationary Schrödinger equation (and taking into account that $\dfrac{d^2}{dx^2} \sin ax = -a^2 \sin ax$), we find:

$$E\, C_n \sin \frac{\pi n x}{L} = \frac{\hbar^2}{2m} \left(\frac{\pi n}{L} \right)^2 C_n \sin \frac{\pi n x}{L},$$

This relation holds if, and only if, the energy $E = \dfrac{\hbar^2}{2m}\left(\dfrac{\pi n}{L}\right)^2$.

This is important; a particle in a box *cannot* have an arbitrary energy. Its energy is *quantized*, taking one of the allowed values:

$$E_n = \frac{\hbar^2}{2m}\left(\frac{\pi n}{L}\right)^2, \quad n = 1, 2, \dots$$

(We start from $n = 1$, because if $n = 0$, our wave function will be identically zero, that is, our box contains no particle).

This quantization is different from the one in a harmonic oscillator; the energy does not change in chunks of $\hbar\omega_0$. Instead the energy difference between the allowed values grows with the number n. The zero-point energy is also different: it is

$E_1 = \dfrac{\hbar^2}{2m}\left(\dfrac{\pi}{L}\right)^2$. But it is energy quantization all the same. A particle in a box has what is called *a discrete energy spectrum*, that is, a set of sharply defined energy levels. We should not be too surprised that it differs from the spectrum of a harmonic oscillator: if a particle was constrained to move around the origin by a continuously growing, but always finite, force of the Hooke's law, here it moves freely inside the box but is sharply reflected by its rigid walls.

Since $E = \dfrac{p^2}{2m}$ and energy is quantized, the momentum must be quantized as well:

$$p_n = \frac{\hbar \pi n}{L},$$

The one slight problem here is that if you calculate the expectation value of the momentum operator $\hat{p}$ in a quantum state $\Psi_n(x) = C_n \sin\dfrac{\pi n x}{L}$, you will find that the corresponding integral is exactly zero:

$$\langle \Psi \,|\, \hat{p} \,|\, \Psi \rangle = |C_n|^2 \int_0^L \sin\frac{\pi n x}{L}\frac{\hbar}{i}\left(\frac{d}{dx}\sin\frac{\pi n x}{L}\right)dx \propto \int_0^L \sin\frac{\pi n x}{L}\cos\frac{\pi n x}{L}dx = 0.$$

This was to be expected, since a particle in a box on average goes nowhere and has zero average velocity and momentum (but, of course, non-zero momentum uncertainty, Δp). Mathematically, this is because the full wave function $\Psi_n(x,t)$ is a *standing* wave.

The situation is somewhat different if instead of a box we put our particle in a ring instead of an interval. Strictly speaking, we cannot do this in one dimension – we need at least two dimensions to make the ends of the interval meet. But instead we can perform a neat mathematical trick: impose *periodic boundary conditions*, that is, imagine that once the particle reaches the right end of the segment, it is instantaneously teleported to its left end, keeping the same velocity. (The same trick works in two, three and more dimensions, though it is somewhat harder to imagine.)

Key idea: Periodic boundary conditions

$$f(x) = f(x + nL)$$
$$n = 1, 2,\ldots$$

$$f(x, y) = f(x + nL, y + qM)$$
$$n, q = 1, 2,\ldots$$

$$y = 0, M, 2M,\ldots \quad x = 0, L, 2L,\ldots$$

$$f(x, y, z) = f(x + nL, y + qM, z + pQ)$$
$$n, q, p = 1, 2,\ldots$$

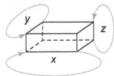

Periodic boundary conditions in one or two dimensions are easy to visualize: you can imagine that your particle 'lives' on a ring or on the surface of a torus. To 'see' it in three dimensions you may need some mental effort.

Mathematically, the trick means that we should look for periodic solutions to the Schrödinger equation, so that $\Psi(x+L) = \Psi(x)$.

You can check directly that two such solutions are

$$\Psi_n^+ = \frac{1}{\sqrt{L}} e^{i(p_n x - E_n t)/\hbar} \text{ and } \Psi_n^- = \frac{1}{\sqrt{L}} e^{i(-p_n x - E_n t)/\hbar}, \text{ if only } p_n = \frac{2\pi n \hbar}{L} = \frac{nh}{L},$$

and, of course, $E_n = \frac{p_n^2}{2m}$.

These states are eigenstates of the momentum operator with the eigenvalues $\pm p_n$:

$$\hat{p}\Psi_n^+ = \frac{\hbar}{i} \frac{d}{dx} \frac{1}{\sqrt{L}} e^{\frac{i(p_n x - E_n t)}{\hbar}} = pn \frac{1}{\sqrt{L}} e^{\frac{i(p_n x - E_n t)}{\hbar}} = p_n \Psi_n^+;$$

$$\hat{p}\Psi_n^- = \frac{\hbar}{i} \frac{d}{dx} \frac{1}{\sqrt{L}} e^{\frac{i(-p_n x - E_n t)}{\hbar}} = -p_n \frac{1}{\sqrt{L}} e^{\frac{i(-p_n x - E_n t)}{\hbar}} = -p_n \Psi_n^-$$

If we calculate the expectation value of the momentum operator in the state Ψ_n^+, we will get $\pm p_n$: these states are *running waves*, which run to the right or to the left.

Let us make two more remarks.

First, since the length of the segment $L = \int dx$, the momentum quantization condition in a ring is nothing else but a version of the Bohr–Sommerfeld quantization condition $\int p\, dq = nh$.

Second, you may ask what happened to the uncertainty principle? After all, the momentum of the particle in the states $\Psi_n^\pm$ is determined, $\Delta p = 0$, while the uncertainty of the position cannot be more than the length of the segment L. The answer is rather subtle. Imposition of periodic boundary conditions on a segment is not *quite* the same as constraining a particle to a ring of the same circumference. In the former case we essentially assume that the particle can travel indefinitely far, but only in a periodic environment (e.g. a crystal), which removes the problem.[7]

---

7   A rigorous treatment of a particle in a ring is, of course, possible, and the uncertainty conditions stay intact, but dealing with it here will cause more trouble than it is worth.

Quantum superposition principle

One of key properties of vectors is that they can be added and subtracted, producing another vector: $A + B = C$. This also holds true in the case of state vectors in a Hilbert space: if $|\Psi_1\rangle$ and $|\Psi_2\rangle$ are state vectors, then $|\Psi_3\rangle = |\Psi_1\rangle + |\Psi_2\rangle$ is also a state vector. (Of course, we would need to normalize this vector before calculating expectation values, but this is a trivial operation.)

We can come to this conclusion from another side as well. Consider the Schrödinger equation. It is a *linear* differential equation, that is, if two functions $\Psi_1(x)$ and $\Psi_2(x)$ are solutions to this equation, then their sum $\Psi_3(x) = \Psi_1(x) + \Psi_2(x)$ is also a solution.

The sum of two state vectors is called their *superposition*, and a simple mathematical fact was given in quantum mechanics a special name: the superposition principle. There is a reason to treat it with so much respect: it does bring about some unusual (from the point of view of classical physics) outcomes.

Let us calculate the expectation value of some observable $\hat{A}$ in a superposition state

$|\Psi_3\rangle = C(|\Psi_1\rangle + |\Psi_2\rangle)$ (the factor C is introduced to make $\|\Psi_3\|^2 = 1$). Clearly,

$$\langle\hat{A}\rangle_3 = C^2\left(\langle\Psi_1|\hat{A}|\Psi_1\rangle + \langle\Psi_2|\hat{A}|\Psi_2\rangle + \langle\Psi_1|\hat{A}|\Psi_2\rangle + \langle\Psi_2|\hat{A}|\Psi_1\rangle\right)$$
$$= C^2\left(\langle\hat{A}\rangle_1 + \langle\hat{A}\rangle_2\right) + C^2\left(\langle\Psi_1|\hat{A}|\Psi_2\rangle + \langle\Psi_2|\hat{A}|\Psi_1\rangle\right).$$

The first bracket does not cause any objections: this is the sum of averages in the states $|\Psi_1\rangle$ and $|\Psi_2\rangle$. For example, if $\hat{A}$ is the position operator, then our particle in state $|\Psi_3\rangle$ on average will be somewhere between its positions in the states $|\Psi_1\rangle$ and $|\Psi_2\rangle$. Unfortunately, there is another bracket, which contains *cross terms* – those with the matrix elements of $\hat{A}$ between the states $|\Psi_1\rangle$ and $|\Psi_2\rangle$. These terms may completely change the pattern. Again, if $\hat{A}$ is the position operator, the presence of these terms can lead to a strange situation when the particle will certainly *not* appear on average anywhere near its positions in the states $|\Psi_1\rangle$ and $|\Psi_2\rangle$.

Key idea: Wave superposition

Wave amplitude = $f(x) + g(x)$

$$+ \; \sqcap\!\sqcup\!\sqcap \; = \; \text{(combined waveform)}$$

Destructive interference

$$+ \; \sqcap\!\sqcup\!\sqcap \; = \; \underline{\qquad}$$

Constructive interference

$$+ \; \sqcap\!\sqcup\!\sqcap \; = \; \sqcap\!\sqcup$$

Superposition of waves: the wave amplitudes add. Therefore two waves with the same wavelength and frequency can cancel or enhance each other depending on their relative phase difference.

For a classical wave this is not strange at all: these *interference patterns* were known and studied extensively in the 19th century. But to learn that this behaviour may be a general feature of Nature and can show up, for example, in particles (it was first observed in electrons), was quite shocking. It was then christened *the particle-wave duality*. This term is being used even now from time to time, though I find it quite unnecessary and rather misleading. It is somewhat like saying that an elephant demonstrates a 'snake-pillar duality'. Quantum systems are what they are: quantum systems, and trying to pin to them purely classical labels is not very productive (Figure 3.2).

Fundamentally, all systems are quantum systems. This book, its author and its reader all consist of quantum particles, which behave as quantum systems do. How, why and under what conditions quantum systems start behaving like classical ones is a very interesting and still not quite resolved problem. We will have a more thorough discussion of this and related questions in Chapters 5, 9, 11 and 12.

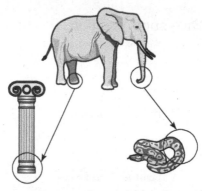

Figure 3.2 Is it reasonable to say that an elephant sometimes behaves as a snake, and sometimes as a pillar?

Quantum statistical mechanics, quantum ensembles and density matrix*

So far we dealt with a *single* quantum system, which was described by a single quantum state vector. They say that such a system is in a *pure* quantum state. All the uncertainties and averages were describing a single system and were the result of the fundamental randomness of Nature, which is reflected in the stochastic character of quantum mechanics – the modern incarnation of the 'Epicurean declinations'.

In reality we often meet with much messier situations. Take for example a cathode ray tube, where electrons, emitted by the cathode, are accelerated by the applied electric field and hit the glass, producing a greenish glow. Each of these electrons is described by a state vector (wave function), which determines the average position of the point where it hits the glass – and the probability that it hits the screen in any other place. If all the electrons were launched in exactly the same quantum state $\Psi(r)$, then the glow would give us an accurate representation of $|\Psi(r)|^2$ in the plane of glass. But this is clearly not so: they are emitted at different points of the cathode, with different energies and momenta, etc. Therefore the glow represents not the square modulus of a single electron's wave function, but an average over many electrons – an ensemble of electrons.

We should therefore generalize our approach to include this extra averaging over the details we cannot (or do not find expedient to) control.

Spotlight: Lev Landau

Lev Davydovich Landau (1908–1968) was a Soviet physicist and Nobel Prize winner (1962).

He introduced the density matrix, created the theory of superfluid helium and made important discoveries in quantum and condensed matter physics. Together with EM Lifshitz, he created the famous multivolume *Course of Theoretical Physics*, translated in several languages and still widely used by physicists all over the world.

According to Landau, there are sciences natural, unnatural and perverse.

This is done (yet again!) in a close resemblance to what we did in classical mechanics. We introduce a *quantum ensemble* – a very large number of copies of our system and a function telling us what fraction of these copies have a given state vector – that is, the probability for our system to be found in the given quantum state (Figure 3.3).

Since we deal with quantum systems, the role of the distribution function is played by an operator – the *statistical operator*, or, as it is called more often, the *density matrix*.

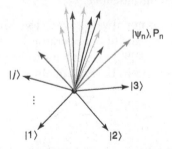

Figure 3.3 A quantum ensemble: the system can be in any of a number of quantum states, with certain probability

Key idea: Trace of a matrix

The trace of a matrix is the sum of its diagonal elements.

$$\hat{M} = \begin{pmatrix} m_{11} & m_{12} & m_{13} & \cdots \\ m_{21} & m_{22} & m_{23} & \cdots \\ m_{31} & m_{32} & m_{33} & \cdots \\ & & \cdots & \end{pmatrix}$$

$$\operatorname{tr} \hat{M} = m_{11} + m_{22} + m_{33} + \ldots$$
$$\operatorname{tr}(\hat{A}\hat{B}) = \operatorname{tr}(\hat{B}\hat{A})$$
$$\operatorname{tr}(\hat{A}\hat{B}\hat{C}) = \operatorname{tr}(\hat{B}\hat{C}\hat{A}) = \operatorname{tr}(\hat{C}\hat{A}\hat{B})$$

To understand how it works, let us once again calculate the expectation value of an observable $\hat{A}$, only now taking into account that we have an ensemble of quantum systems. Accordingly,

$$\langle \hat{A} \rangle = \sum_n P_n \langle \Psi_n | \hat{A} | \Psi_n \rangle.$$

Here we not only average over each quantum state $|\Psi_n\rangle$ from the ensemble, but also over the ensemble itself. The real positive numbers P_n are the probabilities to find the system in the state $|\Psi_n\rangle$. Of course, they add to unity since the system must be certainly (with probability 1) found in *some* state. By the way, if a quantum system can be described by a single state vector, it is said to be in a *pure state*, and if not, then in a *mixed state*, which is quite reasonable: then we deal with a mixture of contributions from the ensemble.

The above formula is not always easy to use. We may already know the matrix elements of our operator $\hat{A}$ in some nice basis, but the ensemble states $|\Psi_n\rangle$ do not have to be (and usually are not) elements of *any* basis. But we can always write any $|\Psi_n\rangle$ as

$$|\Psi_n\rangle = \sum_j C_{nj} |j\rangle,$$

where $|j\rangle$ are the basis vectors. If we substitute such expansions in the formula for $\langle \hat{A} \rangle$, we find:

$$\langle \hat{A} \rangle = \sum_k \sum_j \left(\sum_n C_{nk} P_n C_{nj}^* \right) \langle j|\hat{A}|k \rangle.$$

If we introduce a matrix with elements $\rho_{kj} = \Sigma_n C_{nk} P_n C_{nj}^*$, then

$$\langle \hat{A} \rangle = \sum_k \sum_j \rho_{kj} \langle j|\hat{A}|k \rangle.$$

The matrix (ρ_{kj}) was introduced by Lev Landau and is called the density matrix (or the statistical operator). Now, recalling the rules of matrix multiplication, we see that $\Sigma_j \rho_{kj} \left\langle j \middle| \hat{A} \middle| k \right\rangle = (\hat{\rho}\hat{A})_{kk}$, that is, the kkth (i.e. a diagonal) element of the matrix product of the density matrix and the operator $\hat{A}$ (also written as a matrix in the basis $|j\rangle$, with $A_{jk} = \langle j|\hat{A}|k \rangle$). The expectation value $\langle \hat{A} \rangle$ is then the sum of these elements. In mathematics (linear algebra, to be specific), the sum of diagonal elements of a square matrix is called its *trace*. Therefore we can finally write the expectation value as

$$\langle \hat{A} \rangle = \text{tr}(\hat{\rho}\hat{A}),$$

This expression is more convenient, because the trace has a number of useful properties, which sometimes greatly simplify the calculations. For example, the invariance: no matter in what basis you compute the trace, the result will be the same. Or the cyclic invariance: $\text{tr}(\hat{A}\hat{B}) = \text{tr}(\hat{B}\hat{A}); = \text{tr}(\hat{A}\hat{B}\hat{C}) = \text{tr}(\hat{B}\hat{C}\hat{A}) = \text{tr}(\hat{C}\hat{A}\hat{B})$, and so on.

The Von Neumann equation and the master equation*

For classical probability distribution function, we can use the Liouville equation:

$$\frac{\partial f_N(q,p,t)}{\partial t} = -\sum_j \left\{ \frac{\partial f_N}{\partial q_j} \frac{dq_j}{dt} + \frac{\partial f_N}{\partial p_j} \frac{dp_j}{dt} \right\}.$$

Using Hamilton's equations, we can even rewrite it as

$$\frac{\partial f_N(q,p,t)}{\partial t} = -\sum_j \left\{ \frac{\partial f_N}{\partial q_j}\frac{\partial H}{\partial p_j} - \frac{\partial f_N}{\partial p_j}\frac{\partial H}{\partial q_j} \right\} = -\{f_N, H\},$$

where we have used Poisson brackets. This equation is almost the same as the one for the time derivative for the coordinate or momentum, but the sign is opposite.

If our (or rather Dirac's) earlier insight (that a classical Poisson bracket should transform into a quantum commutator) is right, and the density matrix is indeed a quantum counterpart of the classical distribution function, we should have

$$\frac{\partial f_N(q,p,t)}{\partial t} = -\{f_N, H\} \rightarrow \frac{\partial \hat{\rho}}{\partial t} = -\frac{1}{i\hbar}[\hat{\rho}, \hat{H}].$$

This is actually the case, as can be shown if we write the density matrix as $\hat{\rho} = \sum_n P_n |\Psi_n\rangle\langle\Psi_n|$. You can check if you wish that its matrix elements will be what they should be by taking its time derivative and using the Schrödinger equation for $|\Psi_n\rangle$ and $\langle\Psi_n|$.[8] The equation

$$i\hbar\frac{\partial \hat{\rho}}{\partial t} = -[\hat{\rho}, \hat{H}].$$

is called the *Liouville-von Neumann*, or simply *von Neumann* equation. It looks almost like a Heisenberg equation, but the opposite sign indicates that we are working in the Schrödinger representation. Here the state vectors (and the density matrix) evolve in time, but the operators stay put!

In the Heisenberg representation it is the other way around. The density matrix is constant, the operators depend on time, but the expression for the expectation value,

$$\langle \hat{A}\rangle = \text{tr}(\hat{\rho}\hat{A}),$$

will be the same.

---

8 Mind that $i\hbar\dfrac{\partial\langle\Psi|}{\partial t} = -\langle\Psi|\hat{H}$! Operators act on the bras from the right!

The von Neumann equation would be exact if it included the entire Universe. This is, of course, impossible and unnecessary. After all, we are interested in a particular quantum system and do not care much what happens to everything else. Therefore, instead of the Hamiltonian and the density matrix of the Universe, we modestly use in this equation only the density matrix and the Hamiltonian of the system we are actually interested in. Unfortunately since our system *does* interact with, if not *everything* else, then at least with much of the system's environment, this will not be enough. One has to include in the equation some terms that will describe this interaction. As a result, one obtains the *master equation*

$$i\hbar \frac{\partial \hat{\rho}}{\partial t} = -[\hat{\rho}, \hat{H}] + \Lambda[\hat{\rho}].$$

Here $\Lambda[\hat{\rho}]$ denotes these extra terms. They can be calculated only approximately, but play a key role in describing how a quantum system 'forgets' its past and evolves to an equilibrium state.

Fact-check

1 Heisenberg equations of quantum mechanics describe the evolution of

   a   state vectors
   b   operators
   c   Poisson brackets
   d   density matrix

2 The commutator of two operators is the quantum analogue of

   a   Schrödinger equation
   b   Hamilton function
   c   Rotation
   d   Poisson bracket

3 The Heisenberg and Schrödinger equations of motion are

  a A classical and a quantum equation describing the same system

  b Two different quantum theories

  c Two equivalent ways of presenting the same quantum theory

  d Two early forms of quantum theory superseded by Dirac's theory

4 If the wave function of a quantum particle is a plane wave, it means that

  a It has a definite momentum and no definite position

  b It has a definite position and no definite momentum

  c Both its position and momentum are indefinite and satisfy the Heisenberg uncertainty relation

  d Both position and momentum are definite

5 An operator in Hilbert space can be represented by a

  a square matrix

  b rectangular matrix

  c row vector

  d column vector

6 The matrix product is

  a commutative

  b distributive

  c associative

  d not defined

7 The matrix element of an operator is

  a a real number

  b a column vector

  c a complex number

  d a row vector

8 A pure quantum state is a state of a quantum system, which can be described by

  a an ensemble of quantum state vectors

  b a density matrix

  c a statistical operator

  d a single state vector

9 A mixed quantum state is a state of a quantum system, which can be described by

   a an ensemble of quantum state vectors
   b a density matrix
   c a statistical operator
   d a single state vector

10 The density matrix satisfies the following equation:

   a Liouville
   b von Neumann
   c Liouville-von Neumann
   d Heisenberg

Dig deeper

Further reading. Quantum evolution, Heisenberg and Schrödinger equations.

R. Feynman, R. Leighton and M. Sands, *Feynman Lectures on Physics*, Vols I, II and III. Basic Books, 2010 (there are earlier editions). Vol. III (Chapters 7, 8, 16, 20).

E. H. Wichmann, *Quantum Physics* (Berkeley Physics Course, Volume 4), McGraw-Hill College, 1971. Chapters 6–8.

4

Qubits and pieces

Quantum mechanics is in some respects simpler than classical mechanics. In this chapter we will deal with possibly the simplest object one can encounter in all of exact sciences – a quantum bit.

We have done a lot of heavy lifting in order to explain that the basic structures of quantum mechanics share a lot with those of classical mechanics. It is not our fault that these similarities lie pretty deep and can be seen only using some mathematical tools. But now we can take a breath and apply these tools to simpler quantum objects, which are simpler precisely because they do *not* have classical analogues.

We have seen examples of quantum systems, which have Hilbert spaces with infinitely many dimensions (that is, with infinitely many basis vectors: $\{|j\rangle\}$: $|0\rangle, |1\rangle, |2\rangle, \cdots$). But a quantum system can have as few as two dimensions and such a system is much easier to investigate. (A one-dimensional system would be even simpler, but as such a system can be only in one and the same state all the time, there is nothing to investigate there.)

We will use the example of *quantum bits*, or *qubits*. In principle, any quantum system with a two-dimensional Hilbert space can be a qubit. In practice there are certain additional requirements, which such a system must satisfy. They stem from the intended use of qubits as the basic units of quantum computers, and their physical realization as mostly quite macroscopic devices. But here we will consider the behaviour of ideal qubits.

One important remark before starting: a qubit is actually the simplest physical system one can consistently investigate *from the first principles*, that is, based on the basic physical equations without any simplifications. This seems to contradict our intuition: the behaviour of a particle in space or a harmonic oscillator is much easier to understand. But this ease is the result of a long habit: we are macroscopic beings and are conditioned by evolution to feel at home within classical physics.

What about classical bits – the units of standard computers, which can only be in two states, zero and one? Are not they even simpler? Yes, they are, but they cannot be investigated from the first principles. Classical objects are characterized by a position and momentum, and these can take *any* value whatsoever within some *continuous* range and change *continuously* with time. In order to describe a classical bit switching suddenly between two states, one must make some serious approximations to the initial Hamilton's equations (or their counterparts for the electrical systems).

Case study: Spherical coordinates

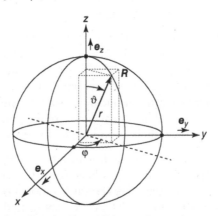

We mentioned earlier that in addition to Cartesian coordinates, one could use other methods of fixing the position of a point in space. One such method is using spherical coordinates when the position of a point with respect to the origin is given, instead of x, y, z, by three other real numbers: the radius r (distance from the origin), the polar angle ϑ (between the vertical direction and the direction to the point; the vertical is identified with the z-axis) and the azimuthal angle φ in the horizontal plane (between the 'east' (x-axis) and the projection of the radius vector on the horizontal plane Oxy).

In other words, in Cartesian coordinates the radius vector can be written as

$$\boldsymbol{R} = r\sin\vartheta\cos\varphi\,\boldsymbol{e}_x + r\sin\vartheta\sin\varphi\,\boldsymbol{e}_y + r\cos\vartheta\,\boldsymbol{e}_z.$$

With little variation, this coordinate system is widely used in astronomy, geography, navigation, anti-aircraft artillery and quantum physics. In geography, for example, the polar angle is called the latitude and is measured not from the North Pole, but from the equator, the azimuthal angle measured from the

Qubits and other two-level quantum systems

Since the mid-1990s, research in the field of quantum computing[1] has attracted special attention to two-level quantum systems. Such systems, aptly called 'qubits' (at once a concatenation of a 'quantum bit' and a homonym of the ancient length unit 'cubit'), were to serve as quantum counterparts to classical bits. From the point of view of purely theoretical quantum computing research, the physical implementation of qubits is irrelevant, which is, of course, not the case when one tries to build a working device.

Key idea: Spin

Spin is the intrinsic property of quantum particles: their 'own' angular momentum, measured in the units of $\hbar$:

$$L = \hbar s.$$

In classical mechanics, a spinning top or a gyroscope has an angular momentum. But in the end, the angular momentum is reduced to the motion of the elements of the spinning top, and, once the top stops spinning, its angular momentum disappears.

On the contrary, the spin angular momentum of quantum particles (for example, electrons) is not reducible to anything else and is their fundamental and unalienable property. (The angular momentum due to the motion of quantum particles in space is called the *orbital* angular momentum.)

If a quantum particle is 'at rest' (that is, its kinetic energy has the minimal value allowed by the uncertainty principle), its spin

---

1 Despite significant progress since the turn of the century, the very possibility of quantum computing on a practically useful scale remains a matter of serious controversy, which can only be resolved by experiment. We will discuss this topic in Chapter 12.

is still present and can be manipulated, controlled and recorded. Spin (mechanical angular momentum) and magnetic moment of a particle are proportional to each other. Therefore we can use the electromagnetic field to perform these operations.

Being a quantum observable, spin is quantized: s can be either integer or half-integer. All the 'substance' particles have half-integer spins:

$$s = \frac{1}{2}, \frac{3}{2} \dots$$

In particular, electrons, protons and neutrons have spin ½. If a particle with nonzero magnetic moment is placed in the constant magnetic field H, the projection of spin s on the direction of the field can have exactly $(2s+1)$ values. Electrons have spin ½, therefore a 'resting' electron (for example, trapped in a potential well) will have $(2 \cdot ½ +1) =$ two spin states and can play the role of a qubit.

Key idea: Electron spin

How small is an electron's spin magnetic dipole moment? The difference in energy between the two states – one, when such magnetic moment is parallel to the magnetic field H, and the other, when it is antiparallel – is $\Delta E = 2\mu_B H$.

Let us take the magnetic field strength 1.5 Tesla, which is approximately 30,000 times the strength of the Earth's magnetic field, or about the same as the field of a powerful magnet in a standard magnetic resonance imaging (MRI) device, which must be approached with caution (it can accelerate ferromagnetic objects – metal clips, etc. – up to a few metres per second, making of them dangerous projectiles). Then $\Delta E \approx 3 \times 10^{-23}$ J. The energy of the falling ant from Chapter 1 is approximately 2×10^{14} (two hundred thousand billion) times greater.

Microscopic two-level quantum systems have been known about since the 1920s and the discovery of the electron spin (intrinsic angular momentum). For example, the spin-related magnetic dipole moment of an electron is approximately one *Bohr magneton* $\mu_B = 9.2740 \dots \times 10^{-24}$ J/T, that is, very small. For a proton it is one *nuclear magneton*,

$\mu_N = 5.0507 \ldots \times 10^{-27}$ J/T, that is, even smaller. Nevertheless this did not stop researchers from using spins as qubits.

One approach was based on using a large number of identical molecules, each containing several atoms with nuclear spins ½ (Figure 4.1) – e.g. ^{1}H, ^{13}C or ^{19}F. These spins interact with each other and could be controlled and measured using nuclear magnetic resonance (NMR) techniques.[2] The enormous number of spins involved compensated the weakness of each spin's signal. Unfortunately, this approach to quantum computing *does not scale*. As one increases the number of qubits (i.e. the spins in a molecule), the signal from them grows relatively weaker, and in order to measure it, the overall number of molecules involved must grow exponentially faster. NMR quantum computers *are not scalable*. For anything practically useful one would have to use an NMR computer of the size of the Sun – not quite an option.

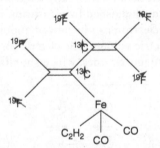

Figure 4.1 NMR quantum computing uses as qubits quantum spins of certain atoms in a molecule. Even though a macroscopic number of molecules are used, the effective number of qubits is the number of 'working' spins in a *single* molecule. The substance in this picture (perfluorobutadienyl iron complex) contains seven qubits and was used in 2001 to perform a real quantum algorithm: factor the number $15 = 3 \times 5$.

This downside of the NMR approach to using all-natural qubits is due to its 'bulk' character, when all spins are controlled and measured at once. In order to overcome it, it is necessary to address each spin individually and in many cases it is simpler

---

2   NMR is discussed later in this chapter.

to use artificial qubits. They have the required parameters (e.g. a stronger coupling to the controlling circuitry) and therefore can be manipulated and measured with more ease. The price to pay is precisely the other side of stronger interaction with the outside world: such interactions tend to disrupt the fragile quantum correlations essential for the operation of a qubit. Nevertheless a very significant progress can be achieved, and by now scientists have realized several different types of artificial qubits and demonstrated that they behave as truly quantum mechanical two-level systems.

One kind of qubit (*charge qubits*) uses quantum dots (Figure 4.2), which are very small islands formed on a solid substrate. Such a dot is so small, that it can contain no more than one extra electron at any one time. (Electrons have the same electric charge and strongly repel each other.) By fabricating two quantum dots next to each other, labelling them '0' and '1' and putting there one electron, we actually make a qubit. The states $|0\rangle$ and $|1\rangle$ can be distinguished by measuring the local electric potential, and the electron can be moved about by applying an appropriate electric voltage. The electric forces are much stronger than the magnetic ones, which simplifies the task.

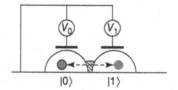

Figure 4.2 A quantum dot-based qubit

If the island is made of a superconducting metal (for example, niobium or aluminium), we obtain a *superconducting charge qubit* (Figure 4.3a). There the difference between the states $|0\rangle$ and $|1\rangle$ is due to the presence or absence of an extra *Cooper pair*.[3]

---

3   In superconducting metals free electrons form correlated pairs – Cooper pairs – which allows them to flow without resistance through the crystal lattice of the material. See Chapter 10.

The advantage of using superconductors is that, due to their special properties, they are better protected from external noise. For example, a loop made of a superconducting wire can carry the electric current forever. A small enough superconducting loop, interrupted by *Josephson junctions*[4], is *a superconducting flux qubit* (Figure 4.3b). In the state |0⟩ the current flows clockwise and in the state |1⟩ counterclockwise. (The 'flux' refers to the magnetic flux, which these currents produce.)

While with charge qubits, the states |0⟩ and |1⟩ differ by a microscopic quantity: the electric charge of one or two electrons; the currents in flux qubits involve a *macroscopic* number of electrons. Nevertheless these qubits behave as quantum two-level systems, which is a very remarkable fact.

There are many more different types of qubits, which were successfully fabricated and tested, but we now have enough examples of quantum two-level systems and can move on to the description of their behaviour.

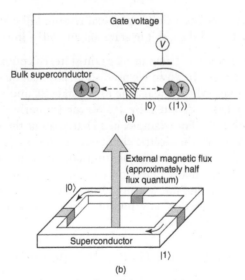

Gate voltage

Bulk superconductor

|0⟩   (|1⟩)

(a)

External magnetic flux (approximately half flux quantum)

|0⟩

Superconductor

|1⟩

(b)

Figure 4.3 Superconducting qubits: (a) charge; (b) flux

---

4   A Josephson junction is a special kind of contact between two pieces of a superconductor. We will discuss the phenomenon of superconductivity in Chapter 10.

Qubit's state vector, Hilbert space, Bloch vector and Bloch sphere

The statement 'the Hilbert space of a qubit is two-dimensional' means simply that any quantum state of a qubit can be written as

$$|\Psi\rangle = \alpha|0\rangle + \beta|1\rangle.$$

Here $|0\rangle$ and $|1\rangle$ are mutually orthogonal[5] unit vectors, and α and β are almost arbitrary complex numbers.

The unit vectors $|0\rangle$ and $|1\rangle$ form the basis of the Hilbert space. Physically they correspond to two distinct quantum states of some physical system, the examples of which we have seen.

The coefficients α and β *are* 'almost' arbitrary, because they must satisfy the condition

$$|\alpha|^2 + |\beta|^2 = 1.$$

This requirement has a simple physical reason: $|\alpha|^2$ is the probability to find the qubit in state $|0\rangle$, and $|\beta|^2$ – in state $|1\rangle$.

'Finding' a quantum system in a certain state, of course, is not completely trivial. To do so we need some physical quantity (an observable), which equals one in the state $|0\rangle$ and zero in the state $|1\rangle$. Let us denote it by $\hat{P}_0$. We can similarly introduce the observable $\hat{P}_1$. (For example, in a charge qubit these observables can be the electric charge (in the units of an electron charge) on the corresponding quantum dot). Measuring one of the observables many times you will on average obtain the probability of finding the system in the corresponding state. Calculating the expectation values of these observables, we see that this is indeed the case:

$$\left\langle \hat{P}_0 \right\rangle = \left\langle \Psi \middle| \hat{P}_0 \middle| \Psi \right\rangle = \left(\alpha^* \langle 0| + \beta^* \langle 1| \right) \hat{P}_0 \left(\alpha|0\rangle + \beta|1\rangle \right) = |\alpha|^2 \, ;$$

$$\left\langle \hat{P}_1 \right\rangle = \left\langle \Psi \middle| \hat{P}_1 \middle| \Psi \right\rangle = \left(\alpha^* \langle 0| + \beta^* \langle 1| \right) \hat{P}_1 \left(\alpha|0\rangle + \beta|1\rangle \right) = |\beta|^2 \, .$$

---

5   That is, with zero scalar product: $\langle 0|1\rangle = \langle 1|0\rangle = 0$.

Geometrically, the requirement $|\alpha|^2 + |\beta|^2 = 1$ simply means that state vector must be normalized, that is

$$\|\Psi\|^2 = 1 = \left(\alpha^* \langle 0| + \beta^* \langle 1|\right)\left(\alpha|0\rangle + \beta|1\rangle\right) = |\alpha|^2 \langle 0|0\rangle + |\beta|^2 \langle 1|1\rangle = |\alpha|^2 + |\beta|^2.$$

This allows a convenient way of writing the two *complex* coefficients α and β in terms of two *real* angles, θ and φ:

$$\alpha = \cos\frac{\theta}{2}; \beta = e^{i\phi}\sin\frac{\theta}{2}.$$

Since $e^{i\phi} \cdot e^{-i\phi} = 1$ and $\sin^2\frac{\theta}{2} + \cos^2\frac{\theta}{2} = 1$, the normalization condition will be automatically satisfied.

The use of real parameters θ and ϕ has an additional advantage: it allows representing the state vector $|\psi\rangle$ of a qubit as a vector in a *three*-dimensional space. (Yes, the Hilbert space of a qubit has only two dimensions, like a plane – but I wish you luck with drawing in a plane a vector whose coordinates are *complex* numbers.)

Consider a vector ρ of unit length starting at origin in the Euclidean space (Figure 4.4). The end of this vector lies on the surface of the unit sphere. Using the spherical coordinates, the vector is fully determined by the two angles, θ (polar) and ϕ (azimuthal) (since $r = 1$):

$$\rho = \sin\theta\cos\phi\, e_x + \sin\theta\sin\phi\, e_y + \cos\theta\, e_z.$$

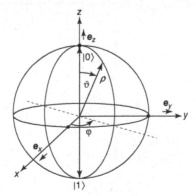

Figure 4.4 Representation of a qubit state vector as a vector (*Bloch vector*) in 3D space

The vector ρ is called the *Bloch vector*. Let us identify these angles with those which characterized our quantum state $|\psi\rangle$, the northern pole of the sphere (where $\theta = 0$ and $\rho = e_z$) with the qubit state $|0\rangle$, and the southern pole where ($\theta = \pi$, and $\rho = -e_z$) with the state $|1\rangle$. Indeed, if $\theta = 0$, from our choice of parameters

$$|\Psi\rangle = \cos\frac{0}{2}\,|0\rangle + e^{i\phi}\sin\frac{0}{2}\,|1\rangle = |0\rangle,$$

while for $\theta = \pi$

$$|\Psi\rangle = \cos\frac{\pi}{2}\,|0\rangle + e^{i\phi}\sin\frac{\pi}{2}\,|1\rangle = e^{i\phi}|1\rangle = |1\rangle.$$

(Since we are at the pole, the azimuthal angle does not matter – any direction from the South Pole is due north!)

What happens when $\theta = \pi/2$? Then our vector ρ lies in the xy-plane:

$$\rho = \sin\frac{\pi}{2}\cos\phi\,e_x + \sin\frac{\pi}{2}\sin\phi\,e_y + \cos\frac{\pi}{2}e_z = \cos\phi\,e_x + \sin\phi\,e_y.$$

Let us draw this plane separately (Figure 4.5). If it reminds you of the complex plane, with the real axis Ox and the imaginary axis Oy, you are absolutely right. We can treat this plane as a complex plane. Indeed, the qubit state vector is now (remember Euler!):

$$|\Psi\rangle = \cos\frac{\pi}{4}|0\rangle + e^{i\phi}\sin\frac{\pi}{4}|1\rangle = \frac{|0\rangle + e^{i\phi}|1\rangle}{\sqrt{2}} = \frac{|0\rangle + (\cos\phi + i\sin\phi)|1\rangle}{\sqrt{2}}.$$

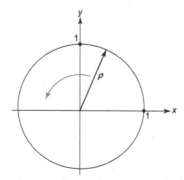

Figure 4.5 Qubit state vector in the xy-plane

Now we see that the azimuthal angle ϕ of the vector ρ is the same, as the angle, which determines the phase of the complex coefficient β in the complex plane. As ϕ increases from zero to 2π, the complex number β makes a counter-clockwise turn about the origin in the complex plane – and so does the vector ρ, turning counter-clockwise about the vertical axis. This is, of course, true not only when ρ lies in the xy-plane (Figure 4.6).

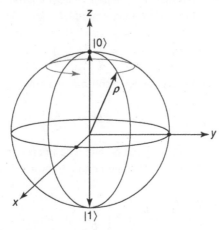

Figure 4.6 The evolution of the Bloch vector ρ

To recapitulate, the quantum state of a qubit (or *any* quantum system, which has only two basis quantum states) can be represented by a unit vector ρ in a three-dimensional space (or, if you wish, by a point on a unit sphere – which is called the *Bloch sphere*). The 'latitude' determines the absolute values of the contributions to the state vector $|\Psi\rangle$ from the states $|0\rangle$ (100% at the North Pole) and $|1\rangle$ (100% at the South Pole). The 'longitude' tells us about the relative phase of these two contributions (that is, the phase φ of the complex coefficient β). The advantage of this approach is that now any abstract evolution of the qubit can be imagined, drawn on a piece of paper, presented on a video clip, and considered using our species' significant visual intuition gained over very long years of evolving, moving, observing, hunting, gathering and manipulating objects in a three-dimensional space.

One counterintuitive feature of this representation is that the orthogonal quantum states are represented not by orthogonal, but by *opposite* Bloch vectors. For example, the states |0) (North Pole) and |1)) (South Pole) are, of course, orthogonal.

Key idea: Transpositions and Hermiticity

An operation, when a matrix is flipped on its diagonal, is called a *transposition* and denoted by a superscript T:

$$\begin{pmatrix} a & b \\ c & d \end{pmatrix}^T = \begin{pmatrix} a & c \\ b & d \end{pmatrix}.$$

The same can be done to a vector column or a row:

$$\begin{pmatrix} a \\ b \end{pmatrix}^T = (a\,b); (a\,b)^T = \begin{pmatrix} a \\ b \end{pmatrix}.$$

The combination of a transposition and a complex conjugation is called the *Hermitian conjugation* and denoted by a superscript + (or a dagger):

$$\begin{pmatrix} a & b \\ c & d \end{pmatrix}^+ = \begin{pmatrix} a^* & c^* \\ b^* & d^* \end{pmatrix};$$

$$\begin{pmatrix} a \\ b \end{pmatrix}^+ = (a^* b^*); (a\,b)^+ = \begin{pmatrix} a^* \\ b^* \end{pmatrix}.$$

For example, a bra vector in the Hilbert space is the Hermitian conjugate of the corresponding ket vector, and vice versa:

$$\left\langle \Psi \right| = \left| \Psi \right\rangle^+ ; \left| \Psi \right\rangle = \left\langle \Psi \right|^+.$$

A matrix that does not change after being Hermitian-conjugated is called a *Hermitian* matrix:

$$\begin{pmatrix} a & b \\ b^* & d \end{pmatrix}^+ = \begin{pmatrix} a & b \\ b* & d \end{pmatrix}.$$

Here numbers a and d must be real. Only Hermitian matrices can represent physical observables.

Qubit Hamiltonian and Schrödinger equation

Since the Hilbert space of a qubit is two-dimensional, any operator of a physical observable – anything that can happen with a qubit and can be measured – is represented by a two-by-two matrix. This is indeed the simplest one can go in physics.

Such small matrices we can draw directly – this is simpler and clearer than a more general form (like (a_{ij})), which we had to use when dealing with infinitely dimensional Hilbert spaces. It is then convenient to represent the state vector as columns (for ket vectors) or rows (for bra vectors) of two numbers:

$$|\Psi\rangle = \alpha|0\rangle + \beta|1\rangle = \left(\begin{array}{c} \alpha \\ \beta \end{array} \right); \langle\Psi| = \alpha^*\langle0| + \beta^*\langle1| = \left(\alpha^* \beta^* \right).$$

We see incidentally that using the standard rule for multiplying matrices

$$\||\Psi\||^2 = \langle\Psi|\Psi\rangle = (\alpha^*\beta^*)\left(\begin{array}{c} \alpha \\ \beta \end{array} \right) = \alpha^*\alpha + \beta^*\beta = |\alpha|^2 + |\beta|^2,$$

as it should be.

Not any two-by-two matrix of complex numbers can represent an operator of a physical quantity (an observable). The key requirement is that this matrix must be *Hermitian*, that is, possess a special kind of symmetry. Its diagonal elements must be real, and off-diagonal ones complex conjugate, that is

$$\hat{A} = \left(\begin{array}{cc} a & c-id \\ c+id & b \end{array} \right),$$

where a, b, c and d are real numbers.

The Hamiltonian of a qubit is usually written in an even simpler form, namely, as

$$\hat{H} = -\frac{1}{2}\begin{pmatrix} \epsilon & \Delta \\ \Delta & -\epsilon \end{pmatrix}.$$

Here ϵ is the *bias*, and Δ is the *tunnelling matrix element*.

Let me explain where this terminology comes from. Imagine a heavy ball, which can roll along a straight line on an uneven surface (for example, along a flexible metal ruler) (Figure 4.7). Due to friction it will eventually settle in one of the local minima, with the potential energy $E_{min} = mgh_{min}$. If there are two such minima close by (forming a *two-well potential*), we can forget about everything else and consider one of them as the state '0' and the other as the state '1' of our system. (We must also assume that the friction is strong enough to quickly stop the ball, so that we could neglect its kinetic energy.) By bending the ruler we can create a bias (that is, an energy difference) ϵ between these states: $\epsilon = E_1 - E_0$.

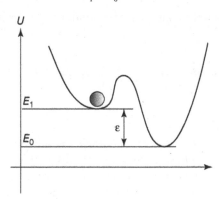

Figure 4.7 Two-well potential

Spotlight: Felix Bloch

Felix Bloch (1905–1983) was a Swiss physicist and Nobel Prize winner (1952) who made significant contributions to quantum physics, in particular the solid state theory. We understand why metals conduct electricity, why insulators do not and why semiconductors have such useful properties, to a large extent due to Bloch. But his Nobel Prize was awarded for something else,

In classical mechanics this would be a model of a classical bit, with one critical flaw: once the ball is in state '0', it will stay there indefinitely, even if make $E_0 > E_1$. In order to switch the system between the states '0' and '1' one would have to take the ball over the barrier between the two potential wells – that is, provide it with enough kinetic energy to sail over it, but not too much, otherwise the ball may not stop in the well '1'. At any rate, an accurate classical description of such a process from the first principles is quite involved.

In quantum mechanics the situation is a bit simpler. If a ball is small enough, and the barrier between the wells low and narrow enough, the ball can *tunnel* between the two wells. We will consider this *quantum tunnelling* phenomenon further in Chapter 6. Here it suffices to say that this process does not require any extra energy and looks like teleportation, which occurs with some average rate Γ. As you recall, frequency times the Planck constant yields energy. The 'energy' $\Delta \sim \hbar\Gamma$ is the *tunnelling* matrix element of the Hamiltonian. It can be calculated for a given system, and the calculation is mostly more straightforward than in the classical case, but here we were only concerned with the origin of this terminology.

Let us now use the qubit Hamiltonian in the Schrödinger equation:

$$i\hbar \frac{d}{dt} \begin{pmatrix} \alpha \\ \beta \end{pmatrix} = -\frac{1}{2} \begin{pmatrix} \epsilon & \Delta \\ \Delta & -\epsilon \end{pmatrix} \begin{pmatrix} \alpha \\ \beta \end{pmatrix} = \begin{pmatrix} -\frac{1}{2}\epsilon\alpha & -\frac{1}{2}\Delta\beta \\ -\frac{1}{2}\Delta\alpha & -\frac{1}{2}\epsilon\beta \end{pmatrix}.$$

This matrix equation is actually a set of two equations: one is the first row, and the other is the second row. If, for example, the qubit is not biased ($\varepsilon = 0$), we obtain

$$i\hbar \frac{d\alpha}{dt} = -\frac{1}{2}\Delta\beta; \; i\hbar \frac{d\beta}{dt} = -\frac{1}{2}\Delta\alpha,$$

or

$$\frac{d\alpha}{dt} = \frac{i\Delta}{2\hbar}\beta; \; \frac{d\beta}{dt} = \frac{i\Delta}{2\hbar}\alpha.$$

Let us start from the state $|0\rangle$. Then initially $\alpha = 1$, $\beta = 0$, and therefore $\frac{d\alpha}{dt} \approx 0; \frac{d\beta}{dt} \approx \frac{i\Delta}{2\hbar}$. This makes sense: the contribution from the state $|1\rangle$ that is, the chance of finding our system in state $|1\rangle$, indeed changes at a rate $\Gamma \sim \frac{\Delta}{\hbar}$. The same, of course, you find if start from the state $|1\rangle$ instead.

Energy eigenstates and quantum beats

In Chapter 3 we looked at *energy eigenstates* – that is, such solutions to the Schrödinger equation, which depend on time via a simple exponent $\exp\left(-\frac{iEt}{\hbar}\right)$ and thus have a definite energy E. They can be found easily enough in case of a qubit.

Such a solution written as $|\Psi_E\rangle = \begin{pmatrix} \alpha \\ \beta \end{pmatrix} \exp\left(-\frac{iEt}{\hbar}\right)$ will satisfy the stationary Schrödinger equation:

$$-\frac{1}{2}\begin{pmatrix} \epsilon & \Delta \\ \Delta & -\epsilon \end{pmatrix}\begin{pmatrix} \alpha \\ \beta \end{pmatrix} = E\begin{pmatrix} \alpha \\ \beta \end{pmatrix}.$$

This system of two equations for two unknown quantities does have a solution only if[6]

$$E = \pm\frac{1}{2}\sqrt{\epsilon^2 + \Delta^2}.$$

---

6 You can trust me on this or consult some textbook on linear algebra.

The lower sign corresponds to the energy E_g of the ground state $|\Psi_g\rangle$ and the upper sign to the energy E_e of the excited state $|\Psi_e\rangle$; see Figure 4.8. The explicit expressions for α and β in these states (with the normalization condition $|\alpha|^2 + |\beta|^2 = 1$) can be also written easily enough, but for us it is sufficient to know that

$$\left|\Psi_g(\infty)\right\rangle = \begin{pmatrix} 1 \\ 0 \end{pmatrix} \text{ and} \left|\Psi_e(\infty)\right\rangle = \begin{pmatrix} 0 \\ 1 \end{pmatrix}, \text{ when } \epsilon \to \infty; \left|\Psi_g(-\infty)\right\rangle = \begin{pmatrix} 0 \\ 1 \end{pmatrix} \text{ and}$$

$$\left|\Psi_e(-\infty)\right\rangle = \begin{pmatrix} 1 \\ 0 \end{pmatrix}, \text{ when } \epsilon \to -\infty; \text{ and} \left|\Psi_g(0)\right\rangle = \frac{1}{\sqrt{2}}\left[\begin{pmatrix} 1 \\ 0 \end{pmatrix} + \begin{pmatrix} 0 \\ 1 \end{pmatrix} \right] \text{ and}$$

$$\left|\Psi_e(0)\right\rangle = \frac{1}{\sqrt{2}}\left[\begin{pmatrix} 1 \\ 0 \end{pmatrix} - \begin{pmatrix} 0 \\ 1 \end{pmatrix} \right], \text{ when } \epsilon = 0.$$

This makes more sense if you look at Figure 4.8. You will see that at a large bias (when the wells are far apart) the state of the system is almost as if it was in either one, or the other of these wells. In other words, the state vector of the system is either $|L\rangle = \begin{pmatrix} 1 \\ 0 \end{pmatrix}$ or $|R\rangle = \begin{pmatrix} 0 \\ 1 \end{pmatrix}$ (calling them 'left' and 'right' is, of course, just a figure of speech). Which of them is the ground, and which the excited state, depends on the sign of the bias.

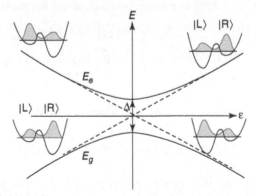

Figure 4.8 Two-well potential: energies and wave functions of the ground and excited qubit states

At zero bias ($\varepsilon = 0$) the energies of the two wells coincide, but the energies of the ground and excited states still differ by Δ. (This is called *level anticrossing*). The energy eigenstates are superpositions of the 'left' and 'right' states.

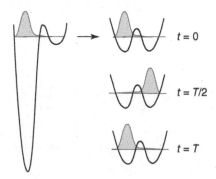

Figure 4.9 Quantum beats

This leads to an interesting phenomenon (Figure 4.9). Suppose we have started with the qubit in state $|L\rangle$ and bias $\varepsilon \to \infty$ (in practice it is enough to bias the qubit and wait: sooner or later because of its interaction with the outside world it will relax to the lowest energy state, which is $|L\rangle$). Let us now very quickly remove the bias. The qubit will be still in the state $|L\rangle$: it did not have time to change it. But this state no longer has a definite energy, because, as we have just said, at $\varepsilon = 0$ such states are $\left|\Psi_g(0,t)\right\rangle = \frac{1}{\sqrt{2}}\big[|L\rangle + |R\rangle\big]\exp\left(\frac{i\Delta t}{2\hbar}\right)$ and $\left|\Psi_e(0,t)\right\rangle = \frac{1}{\sqrt{2}}\big[|L\rangle - |R\rangle\big]\exp\left(-\frac{i\Delta t}{2\hbar}\right)$. (We have restored the time-dependent exponents.)

At $t = 0$ the system is in state $|L\rangle$, which can be written as

$$|L\rangle = \frac{1}{2}\Big[\big(|L\rangle + |R\rangle\big) + \big(|L\rangle - |R\rangle\big)\Big] = \frac{1}{2}\Big[\sqrt{2}\big|\Psi_g(0,t=0)\big\rangle + \sqrt{2}\big|\Psi_e(0,t=0)\big\rangle\Big].$$

What will be the state of the system at any later time? To answer this we only need to put there the time-dependent vectors $|\Psi_g(0,t)\rangle$ and $|\Psi_e(0,t)\rangle$ and obtain

$$|\Psi(t)\rangle = \frac{1}{2}\left[\sqrt{2}\,|\Psi_g(0,t)\rangle + \sqrt{2}\,|\Psi_e(0,t)\rangle\right]$$

$$= \frac{1}{2}\left[|L\rangle\left(e^{\frac{i\Delta t}{2\hbar}} + e^{-\frac{i\Delta t}{2\hbar}}\right) + |R\rangle\left(e^{\frac{i\Delta t}{2\hbar}} - e^{-\frac{i\Delta t}{2\hbar}}\right)\right]$$

$$= \left[|L\rangle\cos\frac{\Delta t}{2\hbar} + i|R\rangle\sin\frac{\Delta t}{2\hbar}\right].$$

This is interesting. Initially ($t = 0$) our qubit was in the 'left' state. We do absolutely nothing with it – but at the time $t = \pi\hbar/\Delta$ it will be in the 'right' state. And it will keep oscillating between these two states. The probability to find the qubit in state $|L\rangle$ or $|R\rangle$ is

$$P_L(t) = \left|\langle L|\Psi(t)\rangle\right|^2 = \left(\cos\frac{\Delta t}{2\hbar}\right)^2 = \frac{1}{2}\left(1 + \cos\frac{\Delta t}{\hbar}\right) \text{ or}$$

$$P_R(t) = \left|\langle R|\Psi(t)\rangle\right|^2 = \left(\sin\frac{\Delta t}{2\hbar}\right)^2 = \frac{1}{2}\left(1 - \cos\frac{\Delta t}{\hbar}\right) \text{ respectively.}$$

This phenomenon is called *quantum beats*. Their period is $T = \frac{2\pi\hbar}{\Delta} = \frac{h}{\Delta}$. Their observation is a good sign that the system we are investigating is indeed a quantum system – that is, it follows the rules of quantum mechanics. It also allows us to directly measure the value of the tunnelling splitting Δ.

Bloch equation, quantum beats (revisited) and qubit control

The time-dependent Schrödinger equation for a qubit can be solved exactly fairly easily, but we will not do it – after all, this is not a quantum mechanics textbook (whatever you may think by now). Instead we go directly to the result. It is has a very simple form, if you use the Bloch vector to represent the quantum state of the qubit (Figure 4.10). It turns out that the Schrödinger equation can be rewritten as the Bloch equation for the Bloch vector:

$$\frac{d}{dt}\rho = \Omega \times \rho.$$

In other words, the vector ρ rotates with the angular velocity $\boldsymbol{\Omega}$,

$$\boldsymbol{\Omega} = \frac{\Delta}{\hbar}\boldsymbol{e}_x + \frac{\in}{\hbar}\boldsymbol{e}_z.$$

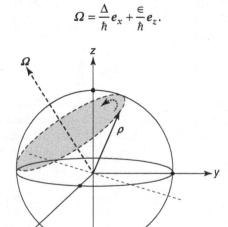

Figure 4.10 Time evolution of the Bloch vector

Key idea: Angular velocity and vector rotation

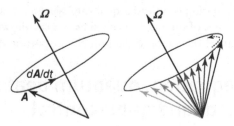

In classical mechanics arbitrary rotations (e.g. of a spinning top) are described with the help of the vector of *angular velocity* $\boldsymbol{\Omega}$. This vector is directed along the axis of rotation (as a right screw) and its length equals the *cyclic* frequency of rotation (so that the period of rotation $T = 2\pi/|\boldsymbol{\Omega}|$). For example, the angular velocity of the Earth is a vector directed approximately from the South to the North Pole and with a length of approximately 7.27×10^{-5} s^{-1}.

(Approximately, because the Earth also rotates around the Sun, and its axis also wobbles slightly because of the imperfectly spherical shape of the globe and the influence of other celestial bodies.)

If a vector $\boldsymbol{A}$ of constant length rotates about its origin with the angular velocity $\boldsymbol{\Omega}$, its time derivative is given by a simple formula, which involves the cross product of two vectors:

$$\frac{d}{dt}\boldsymbol{A} = \Omega \times \boldsymbol{A},$$

that is,

$$\frac{dA_x}{dt} = \Omega_y A_z - \Omega_z A_y;\ \frac{dA_y}{dt} = \Omega_z A_x - \Omega_x A_z;\ \frac{dA_z}{dt} = \Omega_x A_y - \Omega_y A_x.$$

Case study: Cross product of two vectors

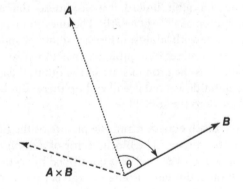

The cross product of two vectors $\boldsymbol{A}$ and $\boldsymbol{B}$ is a vector

$$\boldsymbol{A} \times \boldsymbol{B} = (A_y B_z - A_z B_y)\boldsymbol{e_x} + (A_z B_x - A_x B_z)\boldsymbol{e_y} + (A_x B_y - A_y B_x)\boldsymbol{e_z}.$$

The length of this vector is

$$|\boldsymbol{A} \times \boldsymbol{B}| = AB\sin\Theta,$$

where Θ is the angle between $\boldsymbol{A}$ and $\boldsymbol{B}$ (therefore if two vectors are parallel, their vector product is zero).

Consider, for example, the case of $\varepsilon = 0$ (unbiased qubit). Then the Bloch vector will rotate around the x-axis with the period $T = 2\pi /\Omega = h/\Delta$. If, for example, initially the qubit is in state $|0\rangle$ ($\rho = e_z$), then the Bloch vector will be tracing the zero meridian on the surface of the Bloch sphere. After a half-period we will have $- e_z$, that is, the qubit will be in state $|1\rangle$. The probability of finding the qubit in either state will periodically change between zero and one. This phenomenon is actually the same as *quantum beats*, which we have just previously discussed from a different perspective, and it was observed in all kinds of natural and artificial quantum systems.

This picture helps us to understand how one can control the quantum state of a qubit. Indeed, it is easy to say that a qubit state can be written as $|\Psi\rangle = \alpha|0\rangle + \beta|1\rangle$. The question is, can we *realize* such a state with 'almost arbitrary' α and β? Indeed, in most cases some states of a qubit are easy to realize. If, for example, state $|0\rangle$ is the ground state of the qubit, then we can just cool the qubit down and it will end up there. But how do we get from there to the state $|\Psi\rangle$?

Looking at the Bloch equation and the picture of the Bloch sphere, we see the answer. The Bloch vector of a qubit rotates around the direction of Ω, and Ω is controlled by changing the parameters of the qubit Hamiltonian Δ and ε. (In practice it is usually easier to change ε than Δ, but this is not essential). By changing Ω at certain moments of time, we can make the end of the Bloch vector describe a series of arcs on the Bloch sphere, which will connect any two points on this sphere (Figure 4.11), for example, the points $|0\rangle$ and $|\Psi\rangle = \alpha|0\rangle + \beta|1\rangle$.

Such manipulations are essential for the realization of quantum computing. Of course, sudden changes of the Hamiltonian cannot be realized in practice, so that one uses more or less smooth $\Omega(t)$, and instead of sharply connected arcs the Bloch vector will describe a smooth curve. But the principle of obtaining an arbitrary state of a qubit starting from some other state remains the same.

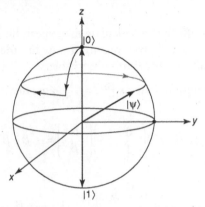

Figure 4.11 Controlling the Quantum state of a qubit

Qubit observables, Pauli matrices and expectation values

Any qubit observable must have the form of a two-by-two Hermitian matrix,

$$\hat{A} = \begin{pmatrix} a & c-id \\ c+id & b \end{pmatrix},$$

where a, b, c and d are all real numbers. Such a matrix can be rewritten as

$$\hat{A} = \begin{pmatrix} \dfrac{a+b}{2} & 0 \\ 0 & \dfrac{a+b}{2} \end{pmatrix} + \begin{pmatrix} \dfrac{a-b}{2} & 0 \\ 0 & -\dfrac{a-b}{2} \end{pmatrix} + \begin{pmatrix} 0 & c \\ c & 0 \end{pmatrix} + \begin{pmatrix} 0 & -id \\ id & 0 \end{pmatrix}$$

$$= \frac{a+b}{2}\begin{pmatrix} 1 & 0 \\ 0 & 1 \end{pmatrix} + \frac{a-b}{2}\begin{pmatrix} 1 & 0 \\ 0 & -1 \end{pmatrix} + c\begin{pmatrix} 0 & 1 \\ 1 & 0 \end{pmatrix} + d\begin{pmatrix} 0 & -i \\ i & 0 \end{pmatrix}.$$

The first of the two-by-two matrices in the second line is the unit matrix, usually denoted by **1**, I or σ_0. The other three are called *Pauli matrices*,

$$\sigma_z = \begin{pmatrix} 1 & 0 \\ 0 & -1 \end{pmatrix}; \sigma_x = \begin{pmatrix} 0 & 1 \\ 1 & 0 \end{pmatrix}; \sigma_y = \begin{pmatrix} 0 & -i \\ i & 0 \end{pmatrix}.$$

It is therefore sufficient to calculate the expectation values of the Pauli matrices in a given quantum state in order to find the expectation value of any qubit observable.

Writing the quantum state as $|\Psi\rangle = \begin{pmatrix} \alpha \\ \beta \end{pmatrix}$, we find:

$$\langle\sigma_x\rangle = (\alpha^*\beta^*) \begin{pmatrix} 0 & 1 \\ 1 & 0 \end{pmatrix} \begin{pmatrix} \alpha \\ \beta \end{pmatrix} = \alpha^*\beta + \beta^*\alpha,$$

$$\langle\sigma_y\rangle = (\alpha^*\beta^*) \begin{pmatrix} 0 & -i \\ i & 0 \end{pmatrix} \begin{pmatrix} \alpha \\ \beta \end{pmatrix} = -i\alpha^*\beta + i\beta^*\alpha,$$

$$\langle\sigma_z\rangle = (\alpha^*\beta^*) \begin{pmatrix} 1 & 0 \\ 0 & -1 \end{pmatrix} \begin{pmatrix} \alpha \\ \beta \end{pmatrix} = |\alpha|^2 - |\beta|^2.$$

Consider, for example, a charge qubit, when in the state $|0\rangle$ the electron is on the left quantum dot, and in the state $|1\rangle$ on the right one. Then the operator of the electric charge on the left dot (in the units of electron charge) can be written as the matrix

$$\hat{Q}_L = \begin{pmatrix} 1 & 0 \\ 0 & 0 \end{pmatrix} = \frac{1}{2} \begin{pmatrix} 1+1 & 0 \\ 0 & -1+1 \end{pmatrix} = \frac{1}{2}(\sigma_z + \sigma_0).$$

Indeed, the expectation value of the electric charge

on the left dot in the state $|\Psi\rangle$ is then given by
$\langle Q_L\rangle = \frac{1}{2}(\langle\sigma_z\rangle + 1) = \frac{1}{2}((|\alpha|^2 - |\beta|^2) + (|\alpha|^2 + |\beta|^2)) = |\alpha|^2$. This is correct, since $|\alpha|^2$ is the probability of finding the electron on the left dot.

Note that since any Hermitian matrix can be written as a combination of the Pauli matrices and the unit matrix, the set $\{\sigma_{x,y,z}, \sigma_0\}$ can be considered as the basis of the 'space' (one more unusual space!) of all two-by-two Hermitian matrices.

Key idea: Pauli matrices

The Pauli matrices greatly simplify the manipulations with qubit observables.
They are all traceless:

$$tr\,\sigma_z = tr\,\sigma_x = tr\sigma_y = 0.$$

You can check that

$$\sigma_x \cdot \sigma_x = 1;\ \sigma_y \cdot \sigma_y = 1;\ \sigma_z \cdot \sigma_z = 1,$$

and that

$$\sigma_x \cdot \sigma_y = i\,\sigma_z;\ \sigma_y \cdot \sigma_z = i\,\sigma_x;\ \sigma_z \cdot \sigma_x = i\,\sigma_y.$$

Qubit density matrix, von Neumann and Bloch equations and NMR*

In reality, a qubit is always influenced by its environment, and even if initially it was in a pure state described by a state vector eventually it will be in a mixed state described by a density matrix. As we know, density matrices are governed by the von Neumann equation, which is analogous to the Liouville equation for the classical probability distribution function:

$$i\hbar\frac{d\hat{\rho}}{dt} = [\hat{H}, \hat{\rho}].$$

The density matrix of a two-level system is a two-by-two Hermitian matrix and can therefore be written as

$$\hat{\rho} = \frac{1}{2}(\rho_x \sigma_x + \rho_y \sigma_y + \rho_z \sigma_z + \sigma_0) = \frac{1}{2}(\boldsymbol{\rho} \cdot \boldsymbol{\sigma} + \sigma_0).$$

The role of the unit matrix is here simple: since all Pauli matrices have trace zero, the unit matrix must ensure that the density matrix has unit trace. All the information about the qubit state is contained in the three coefficients ρ_x, ρ_y, ρ_z, which we have gathered into the single vector $\boldsymbol{\rho}$ (which in the expression for the density matrix is dot-multiplied by the vector $\boldsymbol{\sigma}$ – all this happens in the 'space of all two-by-two Hermitian matrices').

A simple example is the density matrix of a qubit, which is in a pure state described by a state vector $|\Psi\rangle$. It can be written simply as

$$\hat{\rho} = |\Psi\rangle\langle\Psi|,$$

that is,

$$\hat{\rho} = (\alpha|0\rangle + \beta|1\rangle)(\alpha^* \langle 0| + \beta^* \langle 1|) = \begin{pmatrix} |\alpha|^2 & \alpha\beta^* \\ \alpha^*\beta & |\beta|^2 \end{pmatrix}$$

$$= \frac{1}{2}\left\{ (\alpha\beta^* + \alpha^*\beta)\sigma_x + (i[\alpha\beta^* - \alpha^*\beta])\sigma_y + (|\alpha|^2 - |\beta|^2)\sigma_z + (|\alpha|^2 + |\beta|^2)\sigma_0 \right\}.$$

You can check that all the coefficients in the round brackets are always real numbers.

An important point is that the vector ρ in the expression for the density matrix is *the same* Bloch vector, which we introduced earlier in this chapter when describing pure states of a qubit. While the Bloch vectors of pure states had unit length, with their ends on the surface of the Bloch sphere, the ends of the Bloch vectors describing mixed states are *inside* the Bloch sphere (Figure 4.12).

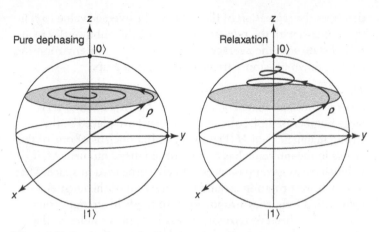

Figure 4.12 Dephasing and relaxation of a qubit state

As before, the Bloch vector of a mixed state will rotate with the angular velocity Ω determined by the Hamiltonian of the qubit, $\frac{d}{dt}\rho = \Omega \times \rho$ But now we can take into account the influence of the environment. The resulting equations are also called the Bloch equations and have the following form (Figure 4.12):

$$\frac{d}{dt}\rho_x = \Omega_y \rho_z - \Omega_z \rho_y - \frac{\rho_x}{T_2};$$

$$\frac{d}{dt}\rho_y = \Omega_z \rho_x - \Omega_x \rho_z - \frac{\rho_y}{T_2};$$

$$\frac{d}{dt}\rho_z = \Omega_x \rho_y - \Omega_y \rho_x - \frac{\rho_z - \langle \rho_z \rangle}{T_1}.$$

These equations give an example of the *master equation* mentioned in Chapter 3. They are not exact, but are simple and give a good approximation, and are therefore very widely used.

The first two terms on the right-hand side of each of these equations, as before, describe the rotation of the Bloch vector. The terms with T_2 describe the *dephasing*: due to them the Bloch vector, spiralling in the horizontal plane, approaches the vertical axis. The energy of the qubit does not change in the process. The *dephasing rate* equals $1/T_2$. Finally, the last equation describes the *relaxation* of the qubit to the average value $\langle \rho_z \rangle$. In this process, with the *relaxation rate $1/T_1$*, the qubit settles to a mixed state with the average energy, dictated by its environment, cooling down or – more frequently – warming up.

The Bloch equations in this form were first worked out in the theory of nuclear magnetic resonance (NMR). There a very large number of molecules (e.g. a patient) are placed in the strong magnetic field *H* (Figure 4.13). If a nucleus of one of the atoms in the molecule has a nonzero magnetic moment *M*, it will behave as a compass needle freely suspended in space. That is, it will start rotating around the direction of the magnetic field with a frequency proportional to the field strength (this rotation is called the *Larmor precession* and is similar to the precession of a spinning top around the vertical, that is, around the direction of the gravitational field).

One can measure the *total* magnetic moment of the sample without much trouble, because it contains a macroscopic number of nuclei. Dividing it by the number of spins involved we obtain the *average* magnetic moment. It obeys the Bloch equations almost identical to those for the Bloch vector:

$$\frac{d}{dt} M_x = \omega_y M_z - \omega_z M_y - \frac{M_x}{T_2};$$

$$\frac{d}{dt} M_y = \omega_z M_x - \omega_x M_z - \frac{M_y}{T_2};$$

$$\frac{d}{dt} M_z = \omega_x M_y - \omega_y M_x - \frac{M_z - \langle M_z \rangle}{T_1}.$$

Here the *Larmor frequency $\omega = \gamma H$* is proportional to the magnetic field and the coefficient γ depends on the type of the nucleus. This dependence is what makes NMR useful. By irradiating the sample with the electromagnetic signal in resonance with the Larmor frequency of a particular nucleus (e.g. of hydrogen ^{1}H) and detecting the response, which is proportional to the average magnetic moment of these nuclei, one can make conclusions about the spatial distribution of this element.

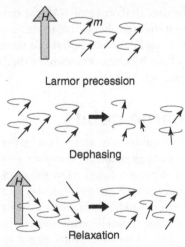

Larmor precession

Dephasing

Relaxation

Figure 4.13 Dephasing and relaxation in NMR

The dephasing and relaxation here have a straightforward physical meaning (Figure 4.13). Dephasing reflects the misalignment of the rotating nuclear magnetic moments: if all of them started in phase (say, pointing at 1 o'clock in the xy-plane), after approximately T_2 they will spread in all directions and the average M_x and M_y will be zero. The relaxation describes the tendency of all magnetic moments to align with the magnetic field (like a compass needle). At absolute zero this is what they would eventually do after the time period T_1. At finite temperature, fluctuations will always nudge the magnetic moment out of alignment and therefore the relaxation will produce some temperature-dependent average magnetic moment $\langle M_z \rangle$.

The similarity between the Bloch equations for the density matrix and for the magnetic moments should not be surprising. Consider, for example, a set of nuclei with spins ½. The total magnetic moment of N such nuclei, NM is the sum of the expectation values of all microscopic magnetic moments:

$$NM = \sum \langle \mu \rangle = N\mathrm{tr}(\hat{\mu}\,\hat{\rho}),$$

where $\hat{\mu} = \kappa\sigma$ is the operator of the magnetic moment, κ is some constant coefficient, and σ has Pauli matrices as its components.

If we calculate the time derivative of *NM*, the only thing that depends on time on the right-hand side of this equation is the density matrix of a single nucleus, $\hat{\rho}$. But this density matrix, if written in terms of the Bloch vector, satisfies the Bloch equations, and so will the quantity *NM*.

Vintage charge qubits

The development of qubits was spurred on by the theoretical discovery that big enough quantum computers (of which qubits should be the unit elements) could crack the so-called public key encryption codes, which are the mainstay of all modern e-commerce and e-communications. Such a device would, of course, be invaluable to intelligence organizations and to organized crime (who – I mean the intelligence agencies – provided some important initial funding for the academic research in this direction), but the goal turned out to be much further away than initially expected (we will discuss this in more detail in Chapter 12), probably for the best of everybody involved. But the progress in designing, making and controlling qubits was all the same very spectacular and would amaze any physicist from, say, the mid-1980s. This progress is described in many long review articles and in a number of books (not to mention the myriad of original research papers), and qubits keep getting better, more reliable and more elaborate as we speak. We will therefore use as the examples some *early* experiments from the turn of this century, which involve *charge* qubits.

We make this choice, first, because these experiments are a bit more straightforward to explain to a non-physicist, and, second, because they were among the first to give the researchers confidence that one can actually observe quantum behaviour in quite large – almost macroscopic – systems.

Let us begin with a superconducting charge qubit (Figure 4.14). This is a small superconducting island ('box') separated from a massive superconductor ('reservoir') by an insulating layer, through which Cooper pairs can penetrate[7] (for some

---

7   The quantum process (tunnelling), which makes this penetration through insulating layers possible, will be discussed in Chapter 6.

important, but not relevant here, reasons in the experimental device the island is attached to the reservoir in two spots). Another insulating layer separates the island from the probe electrode, which is eventually connected to a sensitive ammeter.

The electrodes (bright) are made of aluminium deposited on the insulating substrate (dark). In Figure 4.14 the island, the reservoir and the probe seem to be of one piece, but in reality they are placed on top of each other, with very thin insulating barriers in between.

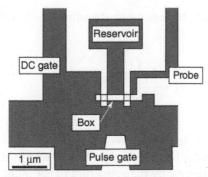

Figure 4.14 Superconducting charge qubit (Nakamura et al., 1999: 786)

The two 'gates' are electrodes, which serve to apply electrostatic voltage to the island. In particular, the 'pulse gate' is connected to equipment, which allows sending very sharp, very precise and very short (just 160 picosecond, that is, 160/1,000,000,000,000 seconds long) voltage pulses.

The whole set is placed inside a *dilution fridge* (a special and expensive kind of refrigerator, which uses liquid helium-3 and helium-4 and cools a tiny bit of its interior to few hundredths of a degree above absolute zero). Aluminium becomes superconductive already at 1.2 degrees above absolute zero (Kelvin), but the system must be cooled further to suppress all the thermal fluctuations to a level when they will no longer disrupt the quantum processes in the qubit.[8]

---

8   This means, in particular, that all the connections between the qubit and the electrical equipment outside the fridge must be very thoroughly engineered to avoid heat and noise entering the fridge.

The island (and therefore its electrical capacitance C) is so small, that placing on it an extra pair of electrons (recall that in a superconductor electrons travel in correlated Cooper pairs) will change its energy by a non-negligible amount. From high-school electrostatics, this energy is $(2e)^2/2C$, where e is the electron charge. In the presence of other electrodes this amount can change, depending on the voltage on these electrodes ('gates'). This allows us to control the energy difference between the states $|0\rangle$ (no extra electrons on the island) and $|1\rangle$ (one extra Cooper pair on the island) by changing the gate voltage.

The Hamiltonian of the superconducting charge qubit has the usual qubit form

$$\hat{H} = -\frac{1}{2}\begin{pmatrix} \epsilon(V_g) & \Delta \\ \Delta & -\epsilon(V_g) \end{pmatrix}.$$

The bias $\epsilon(V_g)$ is directly controlled by the voltage V_g applied to the gate electrodes (dc gate and pulse gate), while the tunnelling term Δ is determined by the properties of the qubit (and can be also tuned by the magnetic field passing through the hole in the 'reservoir' – but this is not important here).

This is the same Hamiltonian we discussed earlier. We take for the 'left' state the state $|0\rangle$, and for the 'right' state the state $|1\rangle$. In the experiment performed by Nakamura, Pashkin and Tsai in 1999 (Figure 4.15) the qubit was initially put in state $|0\rangle$, and the bias ϵ was large. Then a sharp voltage pulse $V_g(t)$ was applied to the gate electrode, such as to make the bias $\epsilon = 0$ for the time duration τ. As we now know, this means that the probability to find the qubit in state $|0\rangle$ (no extra Cooper pair) at that time is $\cos^2(\Delta\tau/2\hbar)$, while the probability to find an extra Cooper pair (state $|1\rangle$) is $\sin^2(\Delta\tau/2\hbar)$ – of course, if our system follows the rules of quantum mechanics.

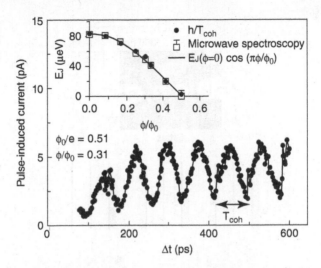

Figure 4.15 Quantum beats in a superconducting charge qubit (Nakamura et al. 1999: 786). Δt here stands for 'τ' in the text.

If the qubit is in state |1⟩ (that is, there is an extra Cooper pair on the island), the two extra electrons will eventually go across the insulating layer into the probe electrode and will be registered by the ammeter as a tiny current pulse. If the qubit is in state |1⟩, then there are no extra electrons on the island and, of course, no current. Thus by measuring the current, one can determine what quantum state the qubit was in at the time of the measurement. In the experiment this cycle was repeated many times every 16 nanoseconds, and the current was measured continuously. Therefore the measured current followed the probability $P_1(\tau)$ to find the qubit in state |1⟩ at the time τ. This measurement produced a single dot on the experimental graph in Figure 4.15.

In order to find the dependence $P_1(\tau)$, the whole scheme had to be repeated for each pulse duration τ. This is a taxing procedure but the result was worth it: the graph resulting clearly shows a sinusoidal dependence of $P_1(\tau)$, that is, the tell-tale quantum beats in a superconducting charge qubit.

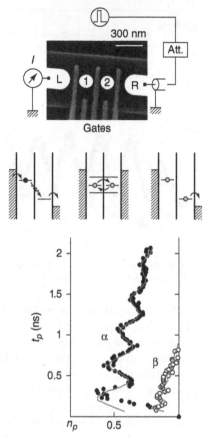

Figure 4.16 Quantum beats in a double-quantum dot charge qubit (from Hayashi et al. 2003: 226804)

A few years later, similar results were obtained for a non-superconducting charge qubit. Hayashi and co-authors in 2003 made a double-quantum-dot charge qubit (Figure 4.16). Here there are *two* quantum dots, so small, that each contains about 25 electrons, and *together* they can contain only one extra electron (bringing in one more would send the electrostatic energy of the system through the roof). By playing with the voltages on gate electrodes (the grey 'comb' in the picture) one can again control whether the state $|0\rangle$ (the extra electron is on the left dot) or the state $|1\rangle$ (the extra electron is on the right dot) has lower energy. (The whole device, of course, had to be placed inside a dilution fridge and cooled down.)

The experiment was essentially the same as with the superconducting charge qubit: starting from the state $|0\rangle$ (extra electron on the left dot) we bring the qubit to the position when the quantum beats begin, wait for a period of time τ, and then measure the current from the right dot. Do this many times. Then change τ and repeat. The quantum beats, obtained after such a procedure, are seen quite clearly though not as well as in the superconducting device. (We have already stated that superconductors have intrinsic advantages as qubit material).

The devices in the pictures are quite small – about a micron across. Nevertheless they already contain a very large number of atoms, while showing a typically quantum behaviour. This was one very important result, supporting the opinion that in quantum mechanics size is not everything: it seems you can be *both* big and quantum.

Fact-check

1 How many quantum spin states does a system with spin 3/2 have?

   a 3

   b 4

   c 8

   d 2

2 Why is an NMR quantum computer not practical?

   a it uses toxic chemicals

   b it uses liquids

   c it works at low temperature

   d it is not scalable

3 The states $|0\rangle$ and $|1\rangle$ of a superconducting charge qubit are different because

    a they carry opposite currents
    b they differ by two electrons
    c they differ by a single electron
    d the electrons are on different quantum dots

4 The Bloch vector represents the quantum state of

    a a superconducting qubit
    b a quantum dot qubit
    c any qubit
    d any quantum system with two-dimensional Hilbert space

5 Two orthogonal quantum states are represented by Bloch vectors, which are

    a orthogonal
    b collinear
    c opposite
    d lying in the same plane

6 If the Hamiltonian of a qubit does not change, the Bloch vector of the qubit will

    a stay still
    b oscillate between two quantum states
    c rotate
    d periodically change its length

7 Quantum beats are the expression of

    a periodic tunnelling between two potential wells
    b rotation of the Bloch vector
    c dephasing
    d oscillation of the Bloch vector

8 Any operator, corresponding to a qubit observable, can be represented using

    a Pauli matrices
    b Pauli matrices and the unit matrix
    c Pauli matrices, the unit matrix and the density matrix
    d Bloch vector and the density matrix

9 NMR uses the following physical process:

    a compression

    b evaporation

    c precession

    d expansion

10 Bloch equations provide an example of:

    a Schrödinger equation

    b Boltzmann equation

    c Liouville equation

    d Master equation

Dig deeper

Further reading I. Quantum mechanics of two-level systems.

R. Feynman, R. Leighton and M. Sands, *Feynman Lectures on Physics*, Vols I, II and III. Basic Books, 2010 (there are earlier editions). Vol. III (Chapters 6, 7, 10, 11).

Further reading II. Quantum bits.

M. Le Bellac, *A Short Introduction to Quantum Information and Quantum Computation*. Cambridge University Press, 2007. Chapter 6.

A. M. Zagoskin, *Quantum Engineering: Theory and Design of Quantum Coherent Structures*, Cambridge University Press, 2011. Chapters 1–3.

5

Observing the observables

In Chapter 2 we discussed observation and measurement. We called the physical quantities, which characterize a quantum system, *observables*. We have stated that they are *operators*, i.e. subtler and more sophisticated mathematical entities than just numbers and vectors, with which we were satisfied in classical physics. We have introduced their *eigenvalues* and their *expectation values* in a given quantum state, that is, what we expect to observe respectively, in a single measurement or on average after many repeated measurements.

All this sounded very natural. Science begins with observation and measurement. But in quantum mechanics these two words acquire a special importance and must be used with precision. Here they both mean the same, namely that measurement (or observation) is *a special kind of interaction* of a quantum system with another system called *a measuring device*, or *apparatus*. The apparatus is supposed to obey the laws of *classical* physics.

Measurement, projection postulate and Born's rule (revisited)

The result of a measurement is that the state vector $|\Psi\rangle$ of the quantum system is *projected* on one of eigenvectors $|a_i\rangle$ of one of the observables $\hat{A}$, which characterize the system. (Which observable it will be depends on the apparatus.) At the same time, the apparatus will evolve from its initial state to some final *classical* state, which is determined by which eigenvector the quantum state was projected to (Figure 5.1). This is called the measurement of the observable $\hat{A}$.

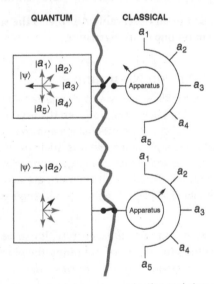

Figure 5.1 Measurement, state vector projection and change in the classical apparatus

One should keep in mind that the observation and measurement in quantum mechanics do *not* imply the existence of an actual – human or alien – sentient observer or even a specially designed automatic measuring device. When an alpha particle hits a crystal of scintillating sphalerite (ZnS(Ag)), it produces a flash of light whether this crystal lies somewhere on the mountainside or is carefully arranged inside a radiation detector. In either case this flash realizes the measurement (or observation) of the particle's position.

The same holds true of course for all hard sciences, including theory of relativity and classical mechanics, but in quantum mechanics with its unusual properties watching one's language is especially important. Keeping this in mind, we can say that the final state of the apparatus, dependent on the final state of the quantum system, $|a_i\rangle$, *reads out* the corresponding eigenvalue a_i of the observable $\hat{A}$.

The *projection postulate*, which is usually associated with the name of von Neumann, is precisely the statement that:

The measurement of an observable $\hat{A}$ *projects* the state of a quantum system on one of its eigenstates.

Spotlight: John von Neumann

John von Neumann (1903–1957) was a Hungarian-American mathematician and physicist. He produced the first mathematically rigorous exposition of quantum mechanics and contributed to the development of nuclear weapons. He was also one of the founders of modern computing and obtained a large number of other exciting and highly influential results in many areas of pure and applied mathematics, physics, economics and engineering.

Note that it does not say *which* eigenstate. It can be *any* eigenstate, and the probability of obtaining the given eigenstate $|a_i\rangle$ is given by the already familiar *Born's rule*:

$$P_i = \left| \langle a_i | \Psi \rangle \right|^2 ,$$

that is, by the square module of the projection of the state vector on the direction of the vector $|a_i\rangle$. The outcome of the measurement is thus totally random.

The measurement, as described by the projection postulate, is thus a rather strange affair. It is not described by the Schrödinger equation: as you recall that was a deterministic equation, which transformed a state vector at one moment of time, $|\Psi(t_1)\rangle$, into a state vector at a later moment of time,

$|\Psi(t_2)\rangle$, without any randomness involved. What we have here instead is that the quantum state vector is turned into a collection (i.e. *ensemble* in the sense of statistical mechanics) of *possible* state vectors, out of which a random choice is made *every time* this measurement of the *same* quantum state vector is repeated. The results of this choice are described by a probability distribution, the set of probabilities P_i. Unlike classical statistical mechanics, though, this randomness is *intrinsic*. It is not due to the complexity of the system, and it cannot be removed by any Laplacian demon, however powerful. This is worth repeating:

> Quantum randomness is a fundamental property of Nature irreducible to any underlying non-random processes.

Mathematically we can say that the measurement transforms a *pure* quantum state $|\Psi\rangle$ into a *mixed* state described by the density matrix $\hat{\rho}$ considered in Chapter 3, which can be mathematically written as [1]

$$\hat{\rho} = \sum_i P_i |a_i\rangle\langle a_i|.$$

The process of such transformation is in many cases virtually instantaneous and completely uncontrollable. This is why it is often called the *collapse of the wave function* (or of the quantum state).

The 'collapse of the wave function', measurement problem and quantum-classical transition

The debates about the meaning of this phenomenon and the seeming contradictions it brings about did long ago spill over

---

1 If the system was in a mixed state to begin with, the outcome will be the same: after a measurement of an observable $\hat{A}$, with certain probability P_i, the system will be found in an eigenstate $|a_i\rangle$ of this observable. The formula for calculating P_i is just slightly more complex than for the initially pure state.

from physics journals to popular literature, science fiction and mainstream media. The general public is, unfortunately, oblivious to the fact that the kind of description of a measurement, which was initially called the 'collapse' (an *instantaneous and uncontrollable* random projection of a quantum state vector on a randomly chosen eigenstate of a given observable; or, if you wish, an *instantaneous and uncontrollable* transformation of a pure quantum state into a specific density matrix) is no longer sufficient. The inertia of many decades of teaching and popularization, a number of old (and excellent!) textbooks using the terminology, and the attractiveness of flashy expressions are, perhaps, the reasons why the term 'collapse' is still being used along with 'particle-wave duality'.

In the current research literature 'collapse of the wave function' or 'collapse of the quantum state', as a rule, means *any* process of transition from a pure quantum state to a mixed state. It does not have to be instantaneous and uncontrollable – actually, it can be made slow and controlled. (But it is still impossible to tell for sure what will be the outcome of a single measurement, i.e. in what eigenstate $|a_i\rangle$ the system ends up after interacting with the apparatus – this is what the density matrix is about, after all.)

The realization did not happen overnight (contrary to the popular beliefs about science, it almost never does). One should recall that when quantum mechanics was being developed, the only known quantum systems were microscopic particles (electrons, protons, alpha-particles), and it was reasonable to believe that any large enough system – like ourselves – will be an apparatus in the sense of quantum measurement. As you have seen in the previous chapter, now we know better: quite large systems, such as superconducting qubits, can and do behave in a thoroughly quantum way. This is not because modern physicists are smarter than the fathers of quantum mechanics (I wish!), but primarily because experimental techniques have made a breath-taking progress since then, and especially over the past two decades. For example, now they allow us to make and handle qubits and what used to be a very speculative research all of a sudden became the matter of practical engineering. This is why the so-called *measurement problem,* along with the intimately related problem of *quantum-classical transition,* was forcefully pushed to the forefront of research.

Measurement problem: explain why and how a quantum state collapses.

Quantum-classical transition: explain why and how the laws of classical mechanics emerge from the laws of quantum mechanics.

The problem of quantum-classical transition is to explain why and how a (presumably large) collection of quantum particles (since elementary particles and atoms are by themselves thoroughly quantum) becomes a classical apparatus. This may seem an example of a sophism the ancient Greeks amused themselves with ('if one pebble is not a pile, and two pebbles are not a pile, then how many pebbles will make a pile?' or 'how many hairs must a man lose to become bald?'), but it is indeed a serious problem deserving a more thorough discussion, which we postpone until Chapter 11. Here we assume that the apparatus is already classical. Then the measurement problem is how to properly describe the process, mechanism and outcomes of the collapse of the quantum state of a quantum system due to its interaction with such an apparatus.

Observing a charge qubit

Let us look again at the superconducting charge qubit (Figure 5.2), which we discussed in Chapter 4. In their experiment, Nakamura, Pashkin and Tsai started with the qubit in the state $|0\rangle$, then applied the gate voltage pulse in such a way, that the qubit state became

$$|\Psi(\tau)\rangle = \cos\frac{\Delta\tau}{2\hbar}|0\rangle + i\sin\frac{\Delta\tau}{2\hbar}|1\rangle.$$

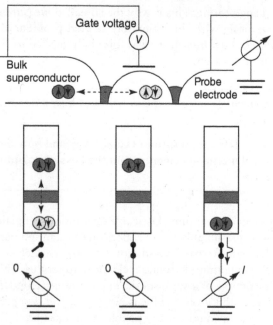

Figure 5.2 Superconducting charge qubit experiment scheme

Following the scheme introduced in Chapter 4, this state vector can be represented by the Bloch vector

$$\rho\left(\tau\right) = \sin\frac{\Delta\tau}{\hbar}e_y + \cos\frac{\Delta\tau}{\hbar}e_z$$

(Figure 5.3), and its evolution – Bloch oscillations – correspond to the rotation of the Bloch vector around the x-axis with the angular frequency $\Delta/\hbar$.[2]

The observation of the qubit state happens when the gate voltage is lifted (at the moment τ) and the extra Cooper pair – if it is there – has the opportunity to leave the superconducting

---

2   Why is the Bloch vector in the yz and not in the xz-plane (or any other plane containing the z-axis)? It is because of the imaginary unit i multiplying the coefficient at $|1\rangle$. Recall that the phase shift between the complex-number coefficients in the expression for the state vector $|\Psi\rangle$ is reflected in the azimuthal angle of the Bloch vector.

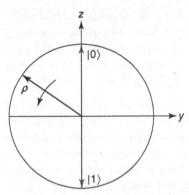

Figure 5.3 Superconducting charge qubit experiment: Bloch vector rotation in *yz*-plane

island into the probe electrode and produce in it a current pulse. This pulse – if it happens – is amplified, transmitted from inside the dilution fridge to another amplifier, and eventually measured and recorded. The probe electrode plays the role of the apparatus; the current pulse, once produced there, moves on in a nice, completely classical manner, like any decent electric current from a wall socket to a tea kettle. The emergence (or non-emergence) of the current pulse is the measurement (observation) of the state of superconducting charge qubit.

Key idea: No 'either/or'

It is wrong to think that the extra two electrons are *either* present *or* absent on the superconducting island at the moment τ. That would be classical, not quantum behaviour. If you *must* use words instead of formulas, you can only say that they are *simultaneously* absent and present on the island until they produce (or do not produce) the current pulse in the probe electrode.

The parameters of the device are chosen in such a way as to make the escape of the extra electrons into the probe fast, and this process is also completely uncontrollable. Therefore it fits the literal understanding of the word 'collapse' pretty well.

The observable that is being measured is the electric charge on the superconducting island (which is, of course, $\hat{Q} = 2e\hat{n}$, where $\hat{n}$ is the *number operator* of extra Cooper pairs on the island). This is because our apparatus is set up to measure the electric current, that is, the charge flow per unit time.

The key consideration here is not the presence of amplifiers and ammeters, but the fact that the large external (with respect to the qubit) system (the probe electrode and connecting wires) 'feels' the extra electric charge on the island. To say this in a more precise and more formal way: if we wanted to write down the Hamiltonian, which would explicitly describe the interaction of the qubit with the probe electrode, this interaction would depend on the charge operator $\hat{Q}$ (or $\hat{n}$, which is essentially the same). This is a *material* fact: if we wanted to measure some other observable, we would have to devise a different apparatus (and if there were no physicists around, the qubit state would collapse all the same – in a way determined by its surroundings). Of course, the electric charge on the island is the easiest observable to observe in our system: this is why it is called a charge qubit.

In terms of the Pauli matrices:

$$\hat{n} = \frac{1}{2}(\sigma_0 - \sigma_z) = \begin{pmatrix} 0 & 0 \\ 0 & 1 \end{pmatrix}.$$

Indeed, the state $|0\rangle = \begin{pmatrix} 1 \\ 0 \end{pmatrix}$ (where there are no extra electrons on the island) is the eigenstate of this operator with the eigenvalue 0:

$$\hat{n}|0\rangle = \begin{pmatrix} 0 & 0 \\ 0 & 1 \end{pmatrix} \cdot \begin{pmatrix} 1 \\ 0 \end{pmatrix} = \begin{pmatrix} 0 \\ 0 \end{pmatrix} = 0|0\rangle,$$

and the state $|1\rangle = \begin{pmatrix} 0 \\ 1 \end{pmatrix}$. is its eigenstate with the eigenvalue 1:

$$\hat{n}|1\rangle = \begin{pmatrix} 0 & 0 \\ 0 & 1 \end{pmatrix} \cdot \begin{pmatrix} 0 \\ 1 \end{pmatrix} = \begin{pmatrix} 0 \\ 1 \end{pmatrix} = 1|1\rangle.$$

After the measurement of the observable $\hat{n}$ the quantum state vector will be projected on one of its eigenstates, either $|0\rangle$ or $|1\rangle$. The corresponding probabilities are

$$P_0 = \left(\cos\tfrac{\Delta\tau}{2\hbar}\right)^2 = \tfrac{1}{2}\left(1 + \cos\tfrac{\Delta\tau}{\hbar}\right) \text{ and } P_1 = \left(\sin\tfrac{\Delta\tau}{2\hbar}\right)^2 = \tfrac{1}{2}\left(1 - \cos\tfrac{\Delta\tau}{\hbar}\right).$$

Obviously, $P_0 + P_1 = 1$, as it should be: we will surely observe *something*.

This kind of measurement (or rather its idealization, when the measurement process is *infinitely* fast) is called **strong (or projective) measurement.**

Bloch vector and density matrix; Bloch equations (revisited); dephasing; and weak continuous measurement*

The result of a single measurement is stochastic. Repeating the measurement of the charge qubit many times in the identical circumstances (in the first place, at the same voltage pulse duration τ), we would still obtain a random series of current pulses, corresponding to the charge of zero and one Cooper pair: e.g. 0, 0, 0, 1, 1, 0, 1..., with zeros and units appearing with relative frequencies P_0 and P_1. Measuring it continuously, we will obtain a signal proportional to P_1. As we know, this is how the experimental curve $P_1(\tau)$ was obtained.

This randomness cannot be included in the (pure) quantum state and is instead incorporated in the density matrix, which in our case is

$$\hat{\rho}(\tau) = P_0(\tau)|0\rangle\langle 0| + P_1(\tau)|1\rangle\langle 1| = \left(\begin{array}{cc} P_0(\tau) & 0 \\ 0 & P_1(\tau) \end{array} \right).$$

The expectation value of charge is, according to what we discussed earlier,

$$\langle \hat{Q} \rangle = 2e \langle \hat{n} \rangle = 2e \, \text{tr}(\hat{\rho}\hat{n}) = 2e \, \text{tr} \left[\begin{pmatrix} P_0(\tau) & 0 \\ 0 & P_1(\tau) \end{pmatrix} \begin{pmatrix} 0 & 0 \\ 0 & 1 \end{pmatrix} \right]$$

$$= 2e \, \text{tr} \begin{pmatrix} 0 & 0 \\ 0 & P_1(\tau) \end{pmatrix} = 2eP_1(\tau).$$

It is instructive to investigate what the strong measurement of a qubit looks like when represented on the Bloch sphere. We start from a pure quantum state $|0\rangle$ that is, with the Bloch vector on the North Pole of the Bloch sphere (Figure 5.4). As we said in Chapter 4, from the Schrödinger equation for the state vector of a qubit it follows, that the Bloch vector ρ will rotate with the angular velocity Ω, determined by the Hamiltonian of the system.

Initially the Hamiltonian was

$$\hat{H}_0 = -\frac{1}{2} \begin{pmatrix} \epsilon & \Delta \\ \Delta & -\epsilon \end{pmatrix} \approx -\frac{1}{2} \begin{pmatrix} \epsilon & 0 \\ 0 & -\epsilon \end{pmatrix}$$

because of the strong bias $|\epsilon| \gg \Delta$ produced by the gate voltages and the corresponding angular velocity $\Omega_0 \approx \frac{\epsilon}{\hbar} e_z$. Since the Bloch vector and the angular velocity are collinear, no rotation takes place – the qubit stays in the state $|0\rangle$.

By suddenly switching the gate voltage the Hamiltonian and the angular velocity were changed to

$$\hat{H}_1 = -\frac{1}{2} \begin{pmatrix} 0 & \Delta \\ \Delta & 0 \end{pmatrix} \text{ and } \Omega_1 \approx \frac{\Delta}{\hbar} e_x.$$

Now the angular velocity is orthogonal to the Bloch vector, and ρ will turn along the meridian, from the North to the South Pole and it will keep turning (that is, undergoing quantum beats) until – at the time τ – the voltage is switched off.

Now the angular velocity once again points along the z-axis. Therefore the Bloch vector will keep turning around it, tracing one of the parallels on the Bloch sphere, but *this* rotation is not important. Indeed, it does *not* change the z-component of the Bloch vector, and this is – loosely speaking – the component we are measuring.

The measurement, that is, the escape (or the non-escape) of the two electrons from the island to the probe electrode, reduces the pure state (with its Bloch vector lying on the surface of the Bloch sphere) to a mixed state (with the Bloch vector inside the Bloch sphere). This density matrix is

$$\hat{\rho}(\tau) = \begin{pmatrix} P_0(\tau) & 0 \\ 0 & P_1(\tau) \end{pmatrix} = \frac{P_0(\tau) + P_1(\tau)}{2}\sigma_0 + \frac{P_0(\tau) - P_1(\tau)}{2}\sigma_z.$$

On the other hand, earlier we wrote it through the Bloch vector as

$\hat{\rho}(\tau) = \frac{1}{2}\left(\sigma_0 + \rho_x\sigma_x + \rho_y\sigma_y + \rho_z\sigma_z\right).$ Therefore the Bloch vector

of our qubit after the measurement will have only z-component:

$$\rho = \frac{P_0(\tau) - P_1(\tau)}{2}e_z = \cos\frac{\Delta\tau}{\hbar}e_z.$$

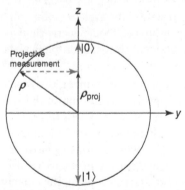

Figure 5.4 Superconducting charge qubit experiment: Bloch vector rotation in *xy*-plane and the projective measurement

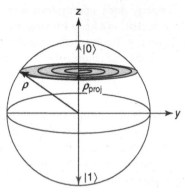

Figure 5.5 Superconducting charge qubit experiment: weak continuous measurement

Key idea: Measurement types

In an ideal ***strong measurement*** of an observable, the state of the system is immediately reduced (*projected*) to one of the eigenvectors of this observable ('collapse of the wave function') and the measurement result is definitely the corresponding eigenvalue of Â.The density matrix describes the *ensemble* of *identical* quantum systems undergoing *identical* measurements.

In an ideal ***weak continuous measurement*** the measurement result only gradually allows us to distinguish between the eigenvalues of Â, and the quantum state of the system evolves continuously. At any moment there is only limited (though increasing) assurance that the result of the measurement corresponds to the state of the system.

In the case of the strong measurement, the Bloch vector instantaneously jumps from the point on the surface of the Bloch sphere to the point on its z-axis (Figure 5.4) and stays there. We have already noted that this is an idealization. It will take some time for the two electrons to move from the island to the probe electrode, and for the probe to 'realize' that the extra charge is there and to start producing the current pulse in the circuitry connected to it. It is therefore plausible to assume that the transition from the surface to the axis of the Bloch sphere will take a finite time.

During this time the Bloch vector will keep rotating about the z-axis, approaching it along a kind of a spiral. This spiral is conveniently and accurately enough described by the Bloch equations in Chapter 4 (Figure 5.5):

$$\frac{d\rho_x}{dt} = \Omega_y \rho_z - \Omega_z \rho_y - \frac{\rho_y}{T_2}; \frac{d\rho_y}{dt} = \Omega_z \rho_x - \Omega_x \rho_z - \frac{\rho_y}{T_2}; \frac{d\rho_z}{dt}$$
$$= \Omega_x \rho_y - \Omega_y \rho_x.$$

The time T_2 characterizes the duration of measurement (though you can see that the measurement described by these equations will, strictly speaking, never end: the spiral will keep winding tighter and tighter, but never reaches the vertical axis). The dephasing time T_2, as we have called it earlier, is a convenient general parameter, which characterizes how fast a pure state is reduced to a mixed state (for whatever reason – not necessarily due to a measurement).[3]

The nice picture of the Bloch vector gradually spiralling towards the axis of the Bloch sphere (and the Bloch equations, which describe this process) indicates the possibility of a slow, rather than instantaneous, measurement of an observable.

Let us once again measure the electric charge $\hat{Q} = 2e\hat{n}$ of a superconducting charge qubit. This time we will not allow the extra Cooper pair to escape to the probe electrode (e.g. by making the insulating barrier between them much thicker) and, instead of measuring the current through the probe electrode, we will measure its electrostatic potential. (This is a silly and impractical scheme for this particular type of qubit, but it will do to explain the basic idea.) The difference made to the electrostatic potential of the probe by an extra Cooper pair on the island will be tiny, and in order to measure it we will have to accumulate the signal over a long period of time.

---

3   Note that the relaxation time T_1 does not appear in the above equation: the measurement by itself does not bring the system to an equilibrium state. In the language of the Bloch vector and Bloch sphere, during the measurement process the Bloch vector will spiral in the same plane, as in Figure 5.6.

Let us denote by Q the value of this accumulated signal, whatever it is. Because our apparatus is now very weakly coupled to the qubit, the reading of Q will be quite uncertain: it will be characterized by a probability distribution $P_Q(q|\Psi;t)$ (the probability of reading out at time t after the measurement started the value of Q equal to q, *if* the qubit is in state $|\Psi\rangle$). The function $P_Q(q|\Psi;t)$ is centred about some *time-dependent* value $q = Q_t(\Psi)$ and has a finite width. At $t = 0$, when no signal is accumulated yet, $Q_0(|0\rangle) = Q_0(|1\rangle)$, and it is impossible to tell whether the qubit is in state $|0\rangle$ or in state $|1\rangle$ (see Figure 5.6).

With time, the width of each distribution grows, but the distance between $Q_t(|0\rangle)$ and $Q_t(|1\rangle)$ grows as well. Fortunately, this distance increases proportionally to t, while the distribution width is only proportional to $\sqrt{t}$, that is, it grows slower. Therefore after some time we will be able to tell with certainty whether the qubit is in the state $|0\rangle$ or $|1\rangle$.

One should remember that initially the qubit could be in *any* state $|\Psi\rangle$. Its 'slow projection' on one of the eigenstates of the charge (or particle number) operator is completely due to its interaction with the apparatus. The *dephasing rate* $1/T_2$ will be exactly the same as the rate Γ at which the apparatus acquires the information about the state of the qubit. (Do not anthropomorphize the apparatus: 'apparatus acquires information' is a shorter way of saying 'due to its interaction with the qubit the classical distribution function of the apparatus changes so that it becomes possible on average to tell which state the qubit was in'.) The dephasing effect of the apparatus on the qubit (or generally on any quantum system that is being measured) is called the detector *back-action*.

This whole exercise illustrates what is known as **weak continuous measurement**. Its final outcome for the quantum system is, of course, the same – its pure state will transform into a mixed state, but the process is now to certain degree controllable. The reason it was not introduced into quantum

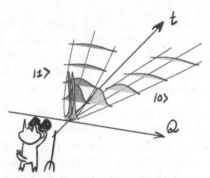

Figure 5.6 Weak continuous measurement (Zagoskin 2011: 266)

mechanics from the outset is simple: at the moment (and for several decades afterwards) all the then experimentally accessible ways of observing the then available quantum systems were very close to a strong projective measurement – so why bother with some boring speculations? When weak measurements became possible, the theory was developed at once.

Measurements and Heisenberg uncertainty relations: Heisenberg microscope

So far we have dealt with measurements of a single observable. Given, it was different from the classical case, because now we cannot tell beforehand *which* eigenvalue of this observable we will actually observe – only the probability of its observation. This is, as we have stated, because the *quantum* state is *not* the just the full set of physical quantities, which characterize it. On the other hand, whatever eigenvalue we get as the result can be measured with an arbitrary precision, just like in classical physics.

Let us now consider the measurement of *two* observables, $\hat{A}$ and $\hat{B}$. Here quantum mechanics immediately seems to get into trouble. The measurement (i.e. the interaction with the apparatus) must reduce the state vector to an eigenvector, but of what operator?

The situation is simple, if $\hat{A}$ and $\hat{B}$ commute, $[\hat{A}, \hat{B}] = 0$. One can prove that these operators will have *common* eigenvectors, so no problem arises. The situation is manifestly different for non-commuting observables and it was explicitly treated by Heisenberg in case of position and momentum.

Heisenberg proposed a *Gedankenexperiment* – a *thought experiment* – when we trace the consequences of some hypothetic situation to their logical end. It is called the *Heisenberg microscope* (Figure 5.7). The idea is simple. Suppose we want to ascertain the position of a particle (say, an electron) by looking at it through a microscope. To do so, we must illuminate the particle. As is known from the classical theory of a microscope, its resolution, that is, the ability to distinguish two points in the object, is of order of the wavelength λ of illuminating light. Therefore the position of the particle can only be found with the uncertainty $\Delta x \sim \lambda$.

In order to increase the accuracy, we must decrease the wavelength of light we are using in the microscope, going from the visible light to ultraviolet, to soft X-rays, hard X-rays, gamma radiation, etc. This is not a problem in a Gedankenexperiment. The problem arises elsewhere.

It is known that light exerts pressure. This prediction of *classical* theory of electromagnetism, made in 1860s, was confirmed in an experiment by Lebedev only in 1900 – the year Planck started it all. The pressure is due to the fact that light is an electromagnetic wave, which interacts with electric charges in the bodies, and it is tiny.

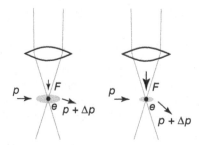

Figure 5.7 Heisenberg microscope

The light pressure is $P = \frac{S}{c}$, where S is the energy flux in the electromagnetic wave (that is, the amount of light energy passing through unit area per unit time), and c is the speed of light. If A is the area of the particle, which scatters light (called *cross-section*), then the light will push the particle with force $F = PA = \frac{SA}{c} = \frac{W}{c}$. Here W is the light power (energy per unit time) scattered by the particle.

Suppose we illuminate the particle for time Δt. Then light will transfer to its so-called *impulse* $F\Delta t = \frac{W\Delta t}{c} = \frac{\Delta E}{c}$, which is proportional to the energy ΔE of light, which was scattered by the particle during the observation. It is known from Newtonian mechanics that an impulse changes the momentum of the body it acted upon: $\Delta p = F\Delta t$. Therefore the *momentum* of the particle will inevitably change after and because of the observation of its *position,* and this change is $\Delta p = F\Delta t = \frac{\Delta E}{c}$.

Can this be avoided? In classical physics certainly: it would be enough to make ΔE arbitrarily small. (In a Gedankenexperiment you do not care about the implementation, only about what is possible *in principle*.) In quantum physics, not so: we have seen that the energy of a quantum oscillator can only change in portions of $\Delta E = h\nu$, and this holds true for light as well. Therefore the minimal change in the particle momentum due to the observation of its position through the Heisenberg microscope is

$$\Delta p = \frac{h\nu}{c} = \frac{h}{\lambda} \sim \frac{h}{\Delta x}.$$

(We have used here the classical relation between the wavelength and the frequency, $\lambda\nu = c$) The better we want to know the position of a particle, the shorter the wavelength must we use, the harder the particle will be hit with the radiation pressure, and the more its momentum will change.[4]

---

4   Turning the argument around, we can say that the Heisenberg microscope demonstrates that photons must carry momentum. This would be a neat theoretical prediction, if light pressure was not predicted and discovered long before.

A more accurate consideration would reproduce the Heisenberg uncertainty relations,

$$\Delta p \Delta x \geq \frac{\hbar}{2}.$$

Our Gedankenexperiment had thus clarified one side of these relations: that the act of measurement (i.e. the interaction with the apparatus – which in this case is the microscope) changes the state of the observed system. This is an explicit example of the detector back-action.

Case study: Gedankenexperiment

Gedankenexperiment, or thought experiment, is a powerful method of reasoning. Though it is best known from the 20th century physics, and associated with such names as Einstein, Bohr and Heisenberg, it has much earlier origins.

Galileo successfully used it in refuting Aristotle's notion that heavier bodies fall faster. He proposed an experiment, where we make a heavy body of two pieces and drop it from a tower (by the way, Galileo himself never intentionally dropped things from towers). Then if the pieces are put together, they should drop faster, while if we insert between them even a very thin sheet of paper, they should drop slower, which is, of course, absurd.

But this is also a dangerous method, since it can produce patently wrong results if used without due care. Even Aristotle made a mistake when he applied it to the existence of vacuum. He reasoned, that in a void a body, once it started to move, will never stop (because in a void all points are the same) and will continue in a straight line. (And this is a perfectly valid conclusion.) Unfortunately, to Aristotle the possibility of such an unlimited rectilinear motion was absurd (as everybody knew in his time, terrestrial bodies fall on earth and stay put, and celestial bodies go in circles). He therefore formulated famous 'Nature abhors vacuum' principle, which was only disproved almost two millennia later.

Standard quantum limit

Suppose now that we want to repeatedly measure the momentum of the same particle. The straightforward (and rather impractical, but we are still in the Gedankenexperiment mode) way of doing it is to measure its position at times $t = 0$ and $t = \Delta t$ calculate

$$p = mv = m(x(t) - x(0))/\Delta t.$$

The first and second position measurements were made with uncertainties Δx_0 and Δx_t. Moreover, because of the first measurement the momentum was changed by at least $\Delta p = \hbar/2\Delta x_0$, and the velocity by $\Delta v = \hbar/2m\Delta x_0$. Therefore this will produce an additional uncertainty of the particle position at time Δt:

$$\Delta x_t{}' = \Delta v \, \Delta t = \frac{\hbar}{2m \, \Delta x_0} \Delta t$$

The three uncertainties, Δx_0, Δx_t and Δx_t, are independent of each other. Therefore by the rules of statistics the uncertainty of the momentum measurement is

$$\Delta p = \frac{m}{\Delta t} \sqrt{(\Delta x_0)^2 + (\Delta x_t)^2 + (\Delta x_t{}')^2}$$

We can take $(\Delta x_t)^2 = 0$; and since the momentum *after* time $t = \Delta t$ is of no interest to us, we are allowed (in our thought experiment) to shine on the particle a beam of light with an arbitrarily small wavelength. But we cannot do so with $(\Delta x_0)^2$: the smaller it is, the greater is the following momentum and velocity perturbation, and therefore $(\Delta x_t{}')^2$.

One can easily check that the expression $x^2 + C/x^2$ is at a minimum, if $x^2 = \sqrt{C}$. Therefore the square root in the expression for Δp will be at a minimum, if

$$(\Delta x_0)^2 = (\Delta x_t{}')^2 = \frac{\hbar \Delta t}{2m}.$$

and the momentum measurement uncertainty in this case is

$$\Delta p_{SQL} = \sqrt{\frac{\hbar m}{\Delta t}}.$$

Note that we have simultaneously measured the average position of the particle,

$$x = \frac{x_0 + x_t}{2}.$$

The uncertainty of this measurement is

$$\Delta x = \frac{1}{2}\sqrt{(\Delta x_0)^2 + (\Delta x_t)^2 + (\Delta x_t')^2},$$

and all the computations we just performed for finding the uncertainty of momentum apply here as well. (No surprise – in this scheme the measurement of momentum was reduced to the measurement of position *and time*). Therefore the minimal uncertainty of position measurement is

$$\Delta x_{SQL} = \frac{1}{2}\sqrt{\frac{\hbar \Delta t}{m}}.$$

SQL stands for *standard quantum limit*. As you see,

$$\Delta x_{SQL} \cdot \Delta p_{SQL} = \hbar / 2.$$

The standard quantum limit indicates the accuracy with which one can usually measure a quantum observable, but it is *not* an absolute limit. It can be sidestepped, as we shall see in a moment.

Quantum non-demolition (QND) measurements and energy-time uncertainty relation

The development of quantum theory of measurement was stimulated by the experimental research in general relativity – a totally unrelated (as one could think) area of physics. The search for the gravitational waves requires very precise measurements of macroscopic systems (e.g. the position of a massive slab of metal, which would be very slightly displaced by a passing gravitational wave), and since the 1960s it was understood that quantum effects would be important. Though no direct observation of gravitational waves was achieved so far,[5] the effect of this research on quantum theory was profound and very fruitful. Yet another confirmation of the old adage: in science, a negative result is still a valuable result.

In precise measurements it is important to determine all possible sources of errors. Assuming that classical noise (such as thermal fluctuations in the system itself and ambient noise reaching it from the outside) is suppressed (by lowering the temperature of the system and properly shielding it), the accuracy seems to be restricted by the standard quantum limit. And it is not.

Here we should be cautious. If we do not know what the quantum state of the system we are going to measure is, then – on average – the expectation value of an observable can be found with an accuracy no better than the standard quantum limit. This was the standard experimental situation for quite a long time after the discovery of quantum mechanics. The situation changed when it became possible to control quantum systems better – in particular, to perform repeated measurements *of the same* observable *in the same* quantum system.

Why was it a game changer? Recall the projection postulate: after the measurement the system ends up in an eigenstate of this observable. Therefore if we immediately measure the same observable *again*, we will *with certainty* obtain *the same* eigenvalue of this observable, with (at least in theory) 100% precision, and the system in question will be found in the *same* quantum state (e.g. $|a_j\rangle$) as before the observation. Therefore this kind of measurement is called a quantum non-demolition (QND) measurement: the state of the system is not 'demolished' and can be observed again and again.

---

5   Their existence is considered proven by observations of a binary pulsar. Its rotation slows down in a good agreement with what is expected from the emission of gravitational waves, which carry away energy and angular momentum of the system.

The fundamental randomness of Nature, expressed by quantum mechanics, somehow allows for this kind of deterministic outcome.

The realization of a QND imposes certain restrictions on the system itself and the apparatus. The instantaneous measurement is an idealization. In reality it will take finite time τ, and it is necessary that the state of the system does not change during this time. Therefore the eigenstate $|a_j\rangle$ of the observable $\hat{A}$, which we measured, must be an eigenstate of the Hamiltonian $\hat{H} = \hat{H}_s + \hat{H}_a$ (here the term $\hat{H}_s$ describes the system itself, and $\hat{H}_a$ – its interaction with the apparatus). This means that not only $\hat{A}|a_j\rangle = a_j|a_j\rangle$ (this is the definition of an eigenstate of $\hat{A}$), but also $(\hat{H}_s + \hat{H}_a)|a_j\rangle = h_j|a_j\rangle$. If substitute the latter expression in the Schrödinger equation, we see that

$$i\hbar\frac{d}{dt}|a_j\rangle = \hat{H}|a_j\rangle = h_j|a_j\rangle.$$

This equation has a simple solution, $|a_j(t)\rangle = \exp(-ih_j t / \hbar)|a_j(0)\rangle$. This means that the direction of the quantum state vector in the Hilbert space is not changed with time – it is only multiplied by a number, $\exp(-ih_j t / \hbar)$, which does not count. (You recall that the probability of observing the system in some state is given by the modulus square of the state vector, which is unaffected by factors like $\exp(-ih_j t / \hbar)$). So we can keep measuring for as long as we like.

The condition that a set of quantum states must be simultaneously eigenstates of an observable $\hat{A}$ and of the Hamiltonian $\hat{H}$ is not easy to satisfy. Mathematically, this is possible only if they commute: $[\hat{A}, \hat{H}] = 0$. As you know, in the Heisenberg representation – when we fix the quantum state and allow the operators to evolve – the Heisenberg equation of motion for the observable $\hat{A}$ is $i\hbar\frac{d}{dt}\hat{A} = [\hat{A}, \hat{H}]$. Therefore the QND condition would be satisfied only by the operators, which – in the Heisenberg representation – do not change in time. Such conserved variables in classical mechanics (for example, the total momentum, total angular momentum or total energy of a closed system) are called the *integrals of motion,* and so they are called in quantum mechanics.

This so-called *strong* QND condition is pretty restrictive. (There is also a *weak* QND condition, which allows non-demolition measurement of other observables as well, but as a price it requires a subtle manipulation with the quantum state, which is at the moment rather impractical.) Nevertheless it is satisfied by some important observables, such as the Hamiltonian itself.

Recall that the time-independent Schrödinger equation, $\hat{H}|\Psi\rangle = E|\Psi\rangle$, clearly says that the eigenvalues of the Hamiltonian are the allowed energies of the system. Obviously, the Hamiltonian commutes with itself, and the strong QND condition is satisfied. Therefore the simplest example of a QND measurement is the measurement of *energy*.

Let us take a quantum particle, freely flying in space. One (rather Gedankenexperiment-esque) way of measuring its energy is to measure its momentum and calculate $E = p^2/2m$ (we do not bother here with relativistic particles). When considering a similar measurement earlier on in the part on 'Standard quantum limit', we assumed $\Delta x_t = 0$, since we did not care that such a precise measurement of the position will change the momentum *afterwards*. Now we can make $\Delta x_t = \Delta x_0$ instead, repeat the calculations and see that the uncertainty of momentum measurement still depends on the duration of measurement as $\sim \sqrt{m\hbar / \Delta t}$ (up to a slightly different numerical factor). Therefore it can be made *arbitrarily small* at the expense of a longer measurement time (and a totally undetermined position of the particle). The energy uncertainty is then

$\Delta E \sim \dfrac{(\Delta p)^2}{2m} = \dfrac{\hbar}{2\Delta t}$, and we come to the *energy–time uncertainty relation:*

$$\Delta E \, \Delta t \geq \frac{\hbar}{2}.$$

This is a very useful relation. For example, it tells how long it will take you (in the very least) to measure energy of some quantum object to the desired precision. And if you want to change the quantum state of a qubit, you cannot go too fast, if you do not want to change the energy of the qubit too much.

Now consider the optical spectrum of an atom. Atoms emit light of definite colours (or invisible radiation of certain frequencies), when they change their quantum state. The bright colours of fireworks are due to such processes in the atoms of sodium (bright yellow), copper (viridian), cadmium (vermillion), etc. The spectral analysis shows their emission lines – that is, the intensity of light as a function of its frequency (or energy, by the Planck formula: $E = h\nu$). These lines have finite width, otherwise we would not see them – and this width $\Delta\nu$ is related to the lifetime τ of the corresponding quantum state (the time that on average it will take the atom to emit light) by the same formula, $\Delta E\,\tau \sim \hbar$, so that $\Delta\nu \sim 1/\tau$. The sharper the line, the longer an atom on average stays in the corresponding quantum state.

An important caveat is to be made here. The status of energy–time uncertainty relation is *different* from that of the bona fide Heisenberg uncertainty relations between conjugate observables. The reason is simple. While the Hamiltonian – the operator of energy – is a perfectly good quantum mechanical observable, **there is no quantum observable of time!** In quantum mechanics time is a parameter, a scalar, a number – and is therefore qualitatively different from position observables, which are all operators. [6] The difference does not invalidate the energy–time uncertainty relation, but sometimes it makes a subtle difference and is worth thinking about.

Quantum Zeno paradox

---

6   This contradicts special relativity with its 'spacetime' – but quantum mechanics does not work for relativistic objects anyway. These are treated by relativistic quantum field theory, which we do not want to wander into. Non-relativistic quantum mechanics is more than enough.

We have already considered strong measurements and weak continuous measurements. Now let us look at a strong continuous measurement. That is, consider a quantum system in which some observable periodically undergoes a strong measurement, and make the intervals between these measurements shorter and shorter. What happens then?

Suppose the system was initially in the eigenstate $|a_0\rangle$ of some observable $\hat{A}$ and we keep strongly observing $\hat{A}$ every Δt seconds. During the time interval Δt the state of the system will evolve, it will become $|a_0\rangle + |\Delta a\rangle$. After the next strong measurement this state of the system will be projected on some eigenstate of $\hat{A}$. It will not necessarily be the same $|a_0\rangle$. Nevertheless, if the time interval Δt, and therefore the deviation $|\Delta a\rangle$, is small enough, it seems plausible that the state will be most likely projected back to the initial eigenstate $|a_0\rangle$.

As a matter of fact, this is exactly what happens. The probability of staying in the same state after one measurement turns out to be

$$P_0(\Delta t) = 1 - \left(\frac{\Delta t}{\hbar}\right)^2 (\delta E_0^2) + \ldots$$

Here (δE_0^2) is the dispersion of energy in the state $|\Delta a_0\rangle$, that is,
$\delta E_0^2 = \langle a_0 | \hat{H}^2 | a_0 \rangle - \left(\langle a_0 | \hat{H} | a_0 \rangle\right)^2$.

(Remember that the system in one of eigenstates of the observable $\hat{A}$ can have a definite energy only if $\hat{A}$ commutes with the Hamiltonian of the system; otherwise $\delta E_0^2 \neq 0$.) After time $t = N\Delta t$ there were N strong measurements, and the probability to remain in the initial state is

$$P_0(t) = \left(1 - \left(\frac{\Delta t}{\hbar}\right)^2 (\delta E_0^2)\right)^N = \left(1 - \left(\frac{\Delta t}{\hbar}\right)^2 (\delta E_0^2)\right)^{\frac{t}{\Delta t}}$$

(the same reasoning tells us that the probability of tossing heads N times in a row is $\frac{1}{2} \times \frac{1}{2} \times \cdots = \left(\frac{1}{2}\right)^N$ – of course if the coin is fair). In the limit of infinitely frequent measurements [7]

$$P_0(t) = \lim_{\Delta t \to 0} \left(1 - \left(\frac{\Delta t}{\hbar}\right)^2 (\delta E_0^2)\right)^{\frac{t}{\Delta t}} = \lim_{\Delta t \to 0} \exp\left(-t\Delta t \frac{\delta E_0^2}{\hbar^2}\right) = 1.$$

Here is a mathematically precise expression of the fact that – at least in quantum mechanics – a watched pot never boils. This could also further justify the use of CCTV – but, unfortunately, if CCTV worked that way, it would stop not just criminal activity, but any activity at all (Figure 5.8).

Figure 5.8 Zeno paradox and CCTV

---

7   Here we used the remarkable formula for the exponent,

$$\lim_{n \to \infty} \left(1 + \frac{z}{n}\right)^n = e^z.$$

Fact-check

1 The measurement requires

   a a measuring device, which obeys the laws of classical physics

   b a measuring device, which obeys the laws of quantum physics

   c an observer, who can manipulate the measuring device

   d a sentient being capable of making an observation

2 The projection postulate states that after the measurement of an observable $\hat{A}$

   a the state of the system will be an eigenstate of $\hat{A}$

   b the state of the system may be an eigenstate of $\hat{A}$

   c the state of the system will instantaneously become an eigenstate of $\hat{A}$

   d the state of the system will be a superposition of eigenstates of $\hat{A}$

3 Quantum randomness is

   a the result of interactions of quantum system with its environment

   b a fundamental property of Nature

   c the expression of our limited knowledge

   d the result of measurement

4 The weak continuous measurement of an observable $\hat{A}$ allows one to

   a determine what will be the outcome of a given measurement

   b control how fast the quantum state is reduced to an unknown beforehand eigenstate of A

   c relax the quantum system to its equilibrium state

   d eliminate quantum uncertainty

5 The evolution of the quantum state during the measurement can be in some cases described by

   a the Schrödinger equation
   b the Liouville-von Neumann equation
   c the instantaneous projection on an eigenstate of an observable
   d the Bloch equations

6 The Heisenberg microscope is a Gedankenexperiment, which

   a illustrates the effect of Heisenberg uncertainty relations on the measuring process of position and momentum
   b provides an argument in favour of photons having momentum
   c shows that it is impossible to measure position or momentum with an arbitrary accuracy
   d reduces the Heisenberg uncertainty relations to the effects of the measuring device

7 Quantum non-demolition measurements

   a allow us to violate the Heisenberg uncertainty relation
   b allow us to measure an observable with an arbitrary accuracy
   c leave the system in the same quantum state as it was before the measurement
   d disrupt the quantum state of the measured system

8 The time-energy uncertainty relation

   a is the result of time and energy being non-commuting operators
   b is different from Heisenberg uncertainty relations
   c follows from time and energy being conjugate variables
   d prohibits the measurement of energy with an arbitrary accuracy

9 The problem of quantum-classical transition is about

    a how the laws of quantum mechanics emerge from the laws of classical mechanics

    b how the laws of classical mechanics emerge from the laws of quantum mechanics

    c what is the source of dephasing

    d how to resolve the internal contradictions of quantum mechanics

10 The quantum Zeno effect means that frequent strong measurements of an observable

    a increase the probability of its leaving the initial quantum state

    b result in the exponential decay of the probability of staying in the initial quantum state

    c freeze it in an arbitrary initial quantum state

    d freeze it in the initial eigenstate of that observable

Dig deeper

Further reading. Measurement and observables.

R. Feynman, R. Leighton and M. Sands, *Feynman Lectures on Physics*, Vols I, II and III. Basic Books, 2010 (there are earlier editions). Vol. III (Chapters 1, 3, 5, 20).

E. H. Wichmann, *Quantum Physics* (Berkeley Physics Course, Volume 4), McGraw-Hill College, 1971. Chapter 6.

A. M. Zagoskin, *Quantum Engineering: Theory and Design of Quantum Coherent Structures*, Cambridge University Press, 2011. Chapter 5.

6

Strange and unusual

We have spent much effort on elucidating both similarities and dissimilarities between quantum and classical mechanics. Some of them are pretty subtle. But the ability of a quantum system to go through solid walls – the *quantum tunnelling* – was certain to attract much attention as soon as it was discovered that it must follow from quantum theory.

We have taken pains to underline the deep similarities between quantum and classical mechanics, in order to better understand the fundamental differences between them. In short, these differences are that:

In classical mechanics the state of the system is essentially identical to the set of physical quantities which characterize the system in this state; they can be considered the coordinates of the state vector.

In quantum mechanics the state and these quantities are separate: the observables are *operators,* which act on the state vector.

In classical mechanics all aspects of a system's evolution are – in principle – fully deterministic.

Quantum mechanics is fundamentally stochastic: the state vector satisfies a fully deterministic Schrödinger equation. Nevertheless the state vector generally does not determine the result of a single measurement (i.e. the outcome of a specific instance of quantum system's interaction with a classical apparatus), but only its probability. Moreover and importantly, this probability is a *nonlinear* (quadratic) function of the state vector.

Spotlight: Ernest Rutherford

Ernest Rutherford (1871–1937), first Baron Rutherford of Nelson, Nobel Prize for Chemistry (1908), was one of the greatest experimental physicists of all times. His main discoveries were in the field of radioactivity. To begin with, he coined the terms 'α and β rays' for the two of prevalent types of radiation. During his work at McGill University in Montreal, Rutherford and Frederick Soddy (1877–1956, Nobel Prize for Chemistry (1921)) discovered that radioactivity is due to *transmutation* of elements (which at the time was considered a fantasy reminiscent of alchemy – atoms were called 'atoms' because of their supposed indivisibility!).

These differences are much less glaring when considered against the common background of (admittedly strange) spaces, where state vectors are evolving. But they inevitably lead to shocking paradoxes, if one tries to imagine quantum objects as very small classical particles (the way Greek atomists and late 19th century physicists and chemists did).

Quantum tunnelling: α-decay

Figure 6.1 Ant in a mug

We will start with *tunnelling*. Let us put a classical particle – say, an ant – in a mug (Figure 6.1). In order to extract it from there we must provide the particle with extra energy, $E = U = mgh$, where m is the mass of the particle, and h is

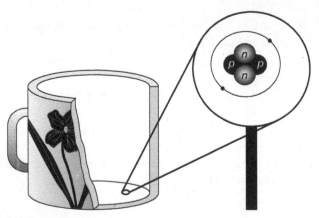

Figure 6.2 A helium atom in a mug

the vertical distance to the mug lip; U is the *height of the potential barrier*, which separates the particle inside the mug from the outside world. For an ant with its weight of 0.5 mg, and taking h =10cm, the height of this barrier is

$$U_{ant} = 0.5 \times 10^{-6} \, \text{kg} \cdot 9.81 \, \text{m}/\text{s}^2 \cdot 0.1 \, \text{m} \approx 0.5 \, \mu\text{J}.$$

If our system is at a finite temperature, it will undergo thermal fluctuations – that is, the energy of the particle can randomly increase (e.g. because the air molecules flying about will hit it in the same direction) – enough for it to clear the mug's lip and escape. Unfortunately, from classical statistical mechanics it follows that the probability of such an event is proportional to $\exp(-U/k_B T)$, where T is the temperature and $k_B = 1.38 \ldots \times 10^{-23}$ J/K is the *Boltzmann constant*. (The factor $\exp(-U/k_B T)$ is called the *Boltzmann*, or the *Gibbs, factor*). Then the probability for the ant to be thrown out of the mug by a thermal fluctuation at a room temperature $(T \approx 300\text{K})$ is about $\exp[-0.5 \times 10^{-6} / (300 \cdot 1.38 \times 10^{-23})] \sim \exp[-10^{14}]$ – and this number is so unimaginably small that I will not even attempt to provide any 'real life' comparisons.

Now replace the ant with an atom of helium-4 (Figure 6.2). It consists of the nucleus (an α-particle, a stable compound of two protons and two neutrons) and two electrons, and its mass is approximately 7×10^{-27} kg. For this atom the probability to escape the mug at 300 K is, as you can easily see, approximately $\exp[-1.7 \times 10^{-6}] \approx 1$. This stands to reason: if such light atoms as helium-4 could not fly at room temperature, nor would much heavier nitrogen or oxygen molecules, and most of the air would just drop on the floor and stay there (forming a 'snow' – this happens on planets far enough from the Sun).

The main point of the comparison is however that either an ant or a helium atom, which escapes the cup, must have energy enough to clear the top of the potential barrier U, so when it will drop back to the ground level it will have at least this much kinetic energy: energy is conserved.

This created a major problem when physicists began seriously thinking about the mechanism of α-radiation. It was known due to Rutherford and Soddy that α-particles are emitted by the nuclei of certain radioactive elements, and that α-particles are the nuclei of helium-4. Positively charged α-particle is strongly repulsed by the positively charged rest of the nucleus and therefore must be held inside by some non-electromagnetic force (so-called *strong* force). Imagine the potential of this force as a 'mug' (properly called a potential well). Its outer side represents the repulsive Coulomb potential, which will accelerate the α-particle once it is out of the nucleus.

The nature of strong force was understood only much later. But whatever it is, if an α-particle escapes from the nucleus, it should have at least enough energy to jump the potential barrier. Then it will accelerate away along the external 'potential slope'. Far enough from the nucleus the repulsive Coulomb potential will become negligible, and the kinetic energy of the α-particle will be equal to its potential energy at the moment it cleared the top (assuming it did so with zero kinetic energy – but this assumption is not actually important).

Spotlight: George Gamow

George Gamow (Georgiy Antonovich Gamov) (1904–1968) was a first-Soviet-then-American physicist, who left his mark in nuclear physics, biology, astrophysics and cosmology. Four young talented Soviet physicists (Gamow, Landau, Ivanenko and Bronshtein)

And here is the rub. If α-particles escaped the nucleus in the same way as helium atoms escape a mug – by randomly acquiring extra energy due to thermal fluctuations – they would have different, random energies. It follows from classical (and quantum) statistical physics that the share of α-particles with energy E would be proportional approximately to the same Boltzmann (or Gibbs) factor $\exp(-E/k_B T)$. Then, the probability of escape over the top is exponentially dependent on the height of the barrier, U, which becomes clear if we rewrite the Gibbs factor as $\exp(-(U + \Delta E)/k_B T) = \exp(-U/k_B T)\exp(-\Delta E/k_B T)$. This probability in turn should be related to the half-life of the radioactive element, which undergoes α-decay: the more the probability, the shorter the lifetime. Therefore one should expect that the shorter the half-life, the lower the energy of emitted α-particles. Finally, one could expect that the half-life depends not only on the energy of α-particles, but also on temperature.

And neither of this actually happens.

All α-particles coming from the same radioactive decay have the *same* energy. For example, the α-particles produced by radium-226 (a popular source of radiation used by Rutherford) have predominantly the energy 4 784.3 keV, while about 5.6% of them have energy 4 601 keV[1] – and nothing in between. Then, an empirical Geiger–Nuttall rule, which quite accurately fits the experimental data, states that the *shorter* the half-life of an α-emitter, the *greater* is the energy of its α-particles, but the dependence is not strongly pronounced. Radioactive elements with as different half-lives as radium-226 (1600 years) and polonium-210 (138 days) emit α-particles with very similar

---

1 One electronvolt is the energy of a single electron charge accelerated by the voltage of 1 volt, that is, $1\,eV \approx 1.6 \times 10^{-19}$ J.

energies (4-5 MeV). Finally, it was established early on that heating or cooling does not affect radioactivity at all.

George Gamow solved the puzzle by applying to it the new quantum mechanics. He considered the Schrödinger equation for a particle in a potential well (Figure 6.3). The external slopes go as *1/r* representing the Coulomb potential, and the internal shape of the well could be for the moment approximated by a vertical wall – producing a shape similar to the nautical mug, which is hard to knock over.

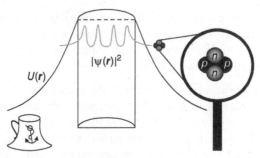

Figure 6.3 α-particle in a potential well

Let us first neglect the outside of the mug and solve the stationary Schrödinger equation inside the well:

$$\left(-\frac{\hbar^2}{2m}\Delta + U(x,y,z) \right)\Psi(x,y,z) = E\,\Psi(x,y,z).$$

It will have only a discrete set of solutions with energies E_1, E_2, E_3, etc., as we have discussed in Chapter 3. But what is crucially important is that these solutions do not immediately go to zero inside the wall, though they drop quite fast – exponentially fast.

Recalling that the square modulus of the wave function gives the probability of finding the particle at a given point, we see that a quantum particle can penetrate inside the potential barrier. (Of course, a classical particle would be just reflected.)

Now you already guessed what happens to α-particles. If the thickness of the mug wall is not too great, the 'tail' of the particle's wave function may stick out a bit – which means a nonzero probability for the α-particle to be found *outside* the nucleus. When this happens, it will be accelerated away by the repulsive Coulomb potential of the rest of the nucleus. And this is what the radioactive α-decay is about. The particle, which does not have enough energy to clear the potential barrier, manages to 'tunnel' right through it. This is *quantum tunnelling*.[2]

This effect nicely explains all the mysterious properties of α-decay.

First, it explains why α-particles from the same decay have the same energy. They had the same quantized energy inside the nucleus – and since they do not change it in the process of tunnelling, they will keep it in the end.

Second, it explains how to reconcile the huge range of half-lives, from fast decaying to stable elements, with a moderate spread of the energies of α-particles. The wave function inside the barrier is $\Psi(x) = C\exp(-k(E)x)$, so if the barrier thickness is say d, then the probability to find the particle outside the barrier will be proportional to $|\Psi(d)|^2 \sim \exp(-2k(E)d)$. This probability directly relates to the half-life of the radioactive element.

The coefficient $k(E) = \sqrt{2m(V - E)/\hbar^2}$, which determines how fast the wave function decays inside the potential wall, does not sharply depend on the energy of an α-particle. But even for the same E this probability – and the half-life – drastically depends on the barrier *thickness, d,* and allows for as broad a variation as you wish.

Finally, the tunnelling mechanism of radioactive decay explained why the nuclei decay at random. This puzzling property of radioactivity turns out to be the result of the fundamental randomness of Nature (or, more specifically, of the Born rule). The wave function determines only the probability of finding an α-particle outside of its nucleusm and cannot tell us more, as a matter of principle.

---

2   This is why off-diagonal matrix elements in the qubit Hamiltonian are called tunnelling matrix elements – they describe the tunnelling of the qubit between two potential minima.

Gamov was able, using this model – that is, a more elaborate and accurate one than the sketch presented here – to explain the Geiger–Nuttall rule and reproduce the experimental data.

Case study: Tunnelling

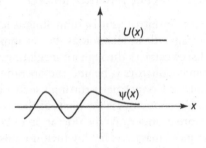

For illustration, let us solve a one-dimensional Schrödinger equation near the wall. Set the wall at $x = 0$, so that at $x < 0$ the potential $U(x) = 0$, and at $x > 0$ it is constant, $U(x) = V$. Then $-\frac{\hbar^2}{2m}\frac{d^2\Psi}{dx^2} = E\Psi$ at $x < 0$ and $\frac{\hbar^2}{2m}\frac{d^2\Psi}{dx^2} + V\Psi = E\Psi$ at $x > 0$. You can check that the first equation has a solution

$$\Psi(x) = A\sin k(E)x + B\cos k(E)x \text{ with } k(E) = \sqrt{\frac{2mE}{\hbar^2}}$$ and the second

one is satisfied by $\Psi(x) = C\exp\left(-k(E)x\right)$ with $k(E) = \sqrt{2m(V-E)/\hbar^2}$.

Here A, B and C are some constants we do not care about at the moment – they are determined from the condition that the wave function 'behaves well' at $x = 0$. As you can see, the wave function indeed does not drop to zero inside the potential barrier, but – of course – the taller the barrier compared to the energy of the particle, the faster it drops.

Quantum tunnelling: scanning tunnelling microscope

Of course, the very idea that an object can just penetrate through a solid wall seemed absurd. Nevertheless, tunnelling is a very widespread phenomenon, but is only happening to small enough objects and low and narrow enough potential barriers. You can amuse yourself estimating the probability that our ant (or even a helium molecule) tunnels out from the real mug at room temperature. You will find that it is much, much smaller than the probability of going over the top for any, but impossibly close to absolute zero (Kelvin).

Strange and counterintuitive as it is, tunnelling is harnessed to play an important role in technology. For example, the tunnelling of charge carriers through an insulating barrier between two semiconductors is behind the operation of tunnel diodes. Tunnelling of Cooper pairs through such a barrier between two superconductors is the basis of the Josephson effect and therefore it underpins the operation of SQUIDs (sensitive detectors of magnetic field, which use this effect and are an indispensable part of, e.g. MRI scanners). It is also crucial for the operation of superconducting qubits, which we have discussed in Chapter 4. Tunnelling of whole atoms or groups of atoms through a potential barrier in a disordered material (which would appear to be a sudden change of a rather large, but still microscopic, section of this material) is the source of a ubiquitous, and hard to get rid of, low frequency noise in electronic devices, so-called 1/f-noise.

Still, the most spectacular use of tunnelling is in scanning tunnelling microscopy – the technology that allowed humans to see atoms for the first time.

The idea of the scanning tunnelling microscope (STM) is deceivingly simple. Consider a solid conducting substrate (sample) probed by a very thin conducting tip at a very short distance d from the sample (Figure 6.4). We assume that both the distance d and the position of the tip in the plane (x, y) above the sample can be precisely controlled (which is the case; of course, one has to fight both the mechanical vibrations of the system and its thermal fluctuations). This is not easy, given that typically $d \sim 10 - 20 \text{Å}$ (that is, 1–2 nm).

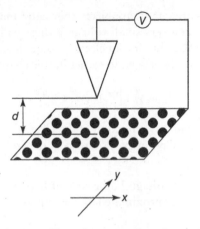

Figure 6.4 Scanning tunnelling microscope: the scheme of operation

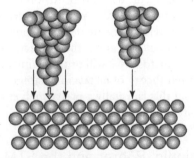

Figure 6.5 The exponential dependence of the tunnelling current on distance increases the resolution of an STM

If electrons were classical particles, they would be unable to 'jump' through the layer of vacuum between the sample and the tip. To make current flow, one would have to apply a strong enough voltage to make possible electron emission from the tip.[3] For smaller voltages, there will be exactly zero current in the circuit.

---

3 We could also heat the system, but this would drastically increase the amplitude of the thermal motion of atoms both in the substrate and in the tip, defeating the purpose of the whole exercise.

Since electrons are quantum objects, they can tunnel across. Therefore even at a very small voltage V there will be some current. The probability for an electron to get from the sample to the tip is, as we have seen in earlier in this chapter,

$$P_E(x, y) = \left| \Psi_E(x, y) \right|^2 e^{-2k(E)d}.$$

Here $\Psi(x, y)$ is the wave function of the electron with energy E in the sample at the point (x, y), and $\kappa(E)$ depends on the energy of the electron.

If we apply a small voltage V, the current between the sample and the tip can be approximated by the expression

$$I(V, x, y) \approx A V N_s(x, y, E_F) e^{-2k(E_F)d}$$

Here A is some coefficient, which depends on the material of the tip; N_s (x, y, E_F) is the so-called *density of electron states* in the sample; and the last term is the tunnelling exponent. The energy E_F is called the *Fermi energy* and is the characteristic energy of electrons in conductors. We will properly introduce it when discussing quantum theory of metals in Chapter 8. For now it is enough to say that this is usually a pretty high energy, about 10 eV (electron Volts)[4] in a typical metal.

Spotlight: Binnig, Rohrer and the STM

Gerd Binnig and Heinrich Rohrer were awarded the Nobel Prize in physics in 1986 for the invention of the scanning tunnelling microscope. This was not the first time that the prize was awarded for an important invention.

Actually prior to 1986, the Nobel Prize in Physics was awarded for inventions in 1908, 1909, 1912, 1927, 1939, 1946, 1953, 1956, 1960, 1964, 1971, and 1978. These prizes ranged from the 1909 Prize awarded to Guglielmo Marconi and Karl Braun 'in recognition of

---

4   One electron Volt (1 eV) is the energy that an electron gains after passing through the voltage difference of 1 Volt. This energy is equivalent to the temperature of approximately 11,600 K. The temperature of the surface of the Sun is only about 6000 K.

their contributions to the development of wireless telegraphy' and the 1912 one to Nils Dalén 'for his invention of automatic regulators for use in conjunction with gas accumulators for illuminating lighthouses and buoys' and to the ones given for the invention of cloud (C.T.R. Wilson 1927) and bubble (Donald Glazer 1960) chambers (at the time, indispensable tools for particle physics); of the cyclotron (E. Lawrence 1939); of the laser (Charles Townes, Nicolay Basov and Aleksandr Prokhorov 1964); and of the transistor (William Shockley, John Bardeen and Walter Brattain 1956).

The density of states is, roughly speaking, the number of electrons with a given energy per unit volume (this is also something we will discuss in more detail in Chapter 8), that is, $N_s(x, y, E_F) \sim |\Psi(x, y, E_F)|^2$. Therefore if we move the tip from point to point and measure the current, the dependence $I(x, y)$ will produce the map of electron density in the sample, that is (since the electron clouds from the 'outer shell' of atoms) the picture of the surface layer of atoms – as simple as that.

Of course, it is not *that* simple. In Figure 6.4 we drew a sharp tip of the probe electrode. But electrodes are made of atoms – and no tip can be made sharper than a single atom (Figure 6.5). Then it seems impossible to 'see' anything smaller than that and the image of atoms on the sample surface should be just a blur.

Fortunately, this reasoning does not take into account the tunnelling exponent in our expression for the current. It is a sharp function of distance the electron has to tunnel. Now, if we look at the 'atomic' picture of the sample and the tip, you will see that the electron will have a much higher probability to tunnel between the parts of the tip and the sample, which are the closest to each other. Therefore the current at a given position of the tip will be dominated by the contribution from the element of the sample surface, which is smaller – maybe significantly smaller – than the size of an atom. Therefore the sample (that is, the electron wave function $|\Psi(x, y)|^2$ on its surface) can be probed with a sub-atomic resolution. A typical STM resolution is 0.1–0.01 nm, while the characteristic radius of an average atom (its external electron shell – which we will talk about in the next chapter) is about 1 nm.

This is a simple theory, and the notebooks of the two Nobel Prize winners contain more or less what we have just presented.

The devil was in the detail of how to make sharp tips (e.g. of gold or tungsten); how to insulate the system from vibrations and protect it from thermal fluctuations; how to realize the precise – very precise – control of the lateral position of the probe in the xy-plane and of the distance between the tip and the sample (too great, and there is no tunnelling and no current; too small, and the tip touches the substrate and is broken). But it was the crucial insight that the exponential dependence of the tunnelling probability on distance would provide a sub-atomic resolution, which justified all this effort.

Now one can buy an STM device off the shelf. These devices are used not just for imaging but also for manipulation of single atoms (Figure 6.6), and not just in physics. Like any important new instrument – like the telescope or the microscope – the STM opened up many new and unexpected venues of research, and all due to quantum mechanical tunnelling and its exponential dependence on distance.

Figure 6.6 The STM tip can not only 'feel', but also move single atoms and place them at the desired spot. Here you see the STM images of 'corrals' built of iron atoms on the (111)-surface of a copper crystal. The ripples are those of the electron density (roughly, the square modulus of the electron wave function). In the case of a round corral the solution is given by a Bessel function (one of mathematical functions akin to sine and cosine). Using an STM is probably the most expensive way of tabulating this function. (a. http://researcher.watson.ibm.com/researcher/files/us-flinte/stm15.jpg b. http://researcher.watson.ibm.com/researcher/files/us-flinte/stm7.jpg

Waves like particles: photoelectric effect and the single-photon double slit experiment

It has been known since the 19th century that light is a wave and that its properties were not quite like the properties of a sound wave. Sound in gases and liquids is a longitudinal wave (that is, the particles oscillate along the same direction in which the wave propagates); sound in solids has both longitudinal and transverse components (in a transverse wave the particles oscillate in the direction, which is orthogonal to the direction of its propagation). But light – and it was established beyond any doubt – had only a transverse component, so it could not be a wave in an ordinary liquid or solid, and the attempts to explain it as a wave in a special substance called 'aether' run into insurmountable difficulties. These difficulties were got rid of by getting rid of aether altogether and recognizing that light (or, more generally, the electromagnetic field) is just a special kind of matter governed by *Maxwell's equations*. In the process the special relativity was built primarily by the efforts of Lorentz, Poincaré and, of course, Einstein (and a very important later contribution from Minkowski), but this is not our concern here.

We have mentioned earlier in Chapter 2 the photoelectric effect. If a metal is illuminated, it may under certain condition emit electrons. At first, nothing strange was seen about this. Electrons are electrically charged, light is an electromagnetic wave, and under the influence of oscillating electric and magnetic fields an electron can absorb enough energy to get loose and escape the metal (where it is held by the collective force of electrostatic attraction of all the negatively charged electrons to all the positively charged ions – see Chapter 8). The flux of emitted electrons – the photoelectric current – was found to be proportional to the light intensity (so-called Stoletov's law, after Aleksandr Stoletov (1839–1896)), as one would expect.

Nevertheless there were features of the phenomenon that refused to fit this picture. To begin with, there was no electron emission if the light frequency was below a certain threshold value ν_0 (different for different materials), no matter how bright the light. Say, a certain metal would emit electrons in a weak green beam, but never when brightly illuminated by a red light.

Then, for light with frequency $\nu > \nu_0$ the electron emission started immediately, even if the light intensity was very low. If the light were acting as a wave, it would take time for an electron to absorb enough energy from a weak wave to be ejected from the metal.

Finally, Philipp Lenard measured the maximal kinetic energy of emitted electrons and found that it is independent on the intensity of light and is determined solely by the frequency of light and the material-dependent threshold frequency (Figure 6.7). In today's notation $E = h(\nu - \nu_0)$.

This certainly could not be explained by any classical wave theory of light.

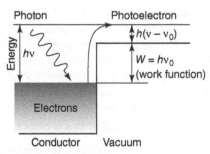

Figure 6.7 Photoelectric effect and Lenard's formula

Key idea: Work functions

Work functions of metals and semimetals are all of the order of few electronvolts, as can be seen in Table 6.1 below.

Table 6.1 Work functions of metals and semimetals

Metal	W eV	ν_0, THz	Light colour
Al	4.1		
Au	5.1		
Be	5.0	1200	ultraviolet
Ca	2.9	700	violet
Cu	4.7		
Fe	4.5		

The threshold frequency and corresponding light wavelength are related to W via

$$\nu_0 = \frac{W}{h}; \lambda_0 = \frac{c}{\nu_0} = \frac{hc}{W}.$$

From here you can easily find the 'red threshold' of the photoelectric effect for a given material and fill in the rest of the table.

The solution to the conundrum was proposed by Einstein (in 1905; it brought him his 1921 Nobel Prize for Physics) based on the then new Planck hypothesis of quanta. Assume that light consists of quanta ('particles', later called photons) with quantized energy $h\nu$. An electron in the metal is hit by a single photon (it would be too improbable for it to be hit by two photons at once), and the maximal energy it can absorb is, of course, $h\nu$. If this energy exceeds some threshold energy (so-called work function) W, this electron can be emitted, and its kinetic energy will be at most $h(\nu - \nu_0)$. Here the threshold frequency $\nu_0 = \frac{W}{h}$. This is the Lenard's observation. And, of course, if $h\nu$ is less than $h\nu_0$, no electron can be emitted at all – which explains the 'red threshold'. Finally, since the intensity of light would be simply proportional to the number of photons falling on the sample per unit time, n, so will be the photoelectric current (number of emitted electrons per unit time) and Stoletov's law is also explained.

This was a neat solution, and well deserving the Nobel Prize. It did not completely settle the question of the quantum nature of light – it took a few more discoveries and Nobel Prizes – but it was a very good start. However it revived the old controversy between Newton and Huygens about the nature of light: is it a flux of particles or a wave? The evidence in favour of the wave character of light was overwhelming – including the massive commercial success of the Carl Zeiss Jena Corporation, which built its formidable reputation in the field of optical devices (from microscopes and telescopes to field binoculars) on wave theory-based recommendations and designs produced by Ernst Abbe (1840–1905), an outstanding physicist and optical scientist. It was therefore assumed that while photons are particles, their interactions with each other, though slight (it was a known fact that light waves do not interact with each other), would produce wave-like behaviour. After all, gases also consist of atoms or molecules, but this does not prevent sound wave propagation.

This stopgap assumption was not consistent for a number of reasons, and was dismissed almost hundred years ago. But it could be shot down in a very spectacular manner, which buries any attempt to keep some 'classical' picture of light, namely single-photon double slit experiments. Students can now perform these experiments (a ready-made kit can be purchased for a few thousand pounds), but this does not subtract from their beauty and deep significance.

The idea of the experiment is simple. If a wave passes through a double slit in a screen, it forms an interference pattern behind the screen, e.g. on a light-sensitive film (Figure 6.8). This happens because each of the slits plays the role of a new source of waves, and the two new waves will now overlap. Thomas Young (1773–1829) used this experiment to demonstrate that light is a wave.

If light consists of particle-like photons (something that would please Newton), but behaves as a wave when there are many of them (this would make Huygens happy), then the double-slit experiment would show it. Send light, which is bright enough – and you will see the interference pattern. Send photons one by one (which now does not present insurmountable technical problems) and there should be no such pattern at all. Instead you should expect just two somewhat blurred lines, created by particles, which flew through either one or the other slit. (Blurred by the finite width of the slits and whatever perturbations the photons could undergo on their way.)

The experiment gave quite different results. When photons were sent one by one, each produced a single spot in the picture – confirming that a photon is a particle. But these spots did not form two blurred lines – instead they produced an interference pattern, as if each single photon was a wave in and of itself and passed through both slits at once (Figure 6.8).

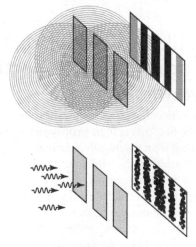

Figure 6.8 A scheme of the double slit experiment

Geoffrey Ingram Taylor (1886–1975) in 1909 conducted for the first time an experiment of this type. Taylor used a 'feeble light', produced by a gas flame shaded by smoked glass screens, to take a picture of a needle. Taken with 'classical' – wave-like – light, such a picture would show dark and light fringes near the edge of the needle's shadow, the same in origin as the interference pattern from a double slit. The longest exposure was '2000 hours or about three months', which corresponded to the apparatus containing on average less than a single photon at a time. Nevertheless, 'in no case was there any diminution in the sharpness of the pattern'. So, even a single photon behaved both as a wave and a particle (of course, the word 'photon' was not invented yet, and even the word 'quantum' was never used in the paper).[5]

---

5   G. I. Taylor went on to make a brilliant career in physics, primarily fluid dynamics and mechanics of deformable solids (which remain some of the most difficult areas of physics). He never returned to quantum mechanics, perhaps finding it not challenging enough.

As we now know well, a photon is neither a particle nor a wave. It is a quantum object described by a quantum operator and quantum state vector. It is an elephant, and not a combination of a snake with some pillars.

Particles like waves: diffraction and interference of massive particles; de Broglie wavelength

Light is not the easiest object of application for quantum mechanics. In the first place, light propagates with the speed of light, which makes it an essentially relativistic object. It must be treated in full accordance with special relativity, where space and time unite to form space-time. As you can immediately see from the very form of the Schrödinger equation, it does not satisfy this requirement: it contains a first-order derivative with respect to time, but second-order derivatives with respect to spatial coordinates. Then, photons are massless. This is why they can travel at the speed of light. This is also why they can be easily produced and eliminated, so that their number is not conserved. Simple quantum mechanics cannot deal with this situation. To handle light one needs from the very beginning a *relativistic quantum field theory*. Such a theory (*quantum electrodynamics*) was built in due time, and it is arguably the most precise theory known to man so far. But it took time, and meanwhile the focus was on quantum mechanics of massive particles – such as electrons, protons, neutrons and α-particles.

The state of such particles can be described by a wave function $\Psi(x, y, z, t)$, which satisfies the Schrödinger equation. What happens if such a particle hits a screen with two slits in it?

The answer is given by the solution of the Schrödinger equation. Consider for simplicity a stationary equation for a particle with energy E and mass m, in the absence of external forces:

$$-\frac{\hbar^2}{2m}\left(\frac{\partial^2}{\partial x^2} + \frac{\partial^2}{\partial y^2} + \frac{\partial^2}{\partial z^2}\right)\Psi(x, y, z) = E\Psi(x, y, z).$$

We must find its solution with appropriate boundary conditions. It is reasonable to demand that the wave function is zero on and inside the impenetrable screen (the particle has zero probability to get there, so $|\Psi|^2$ should be zero as well). The wave function before the screen should look like a combination of incoming and reflected waves. This equation is exactly like a wave equation from the classical theory of light and the mathematical methods for solving such equations were worked out long ago – we have already mentioned this in Chapter 3.[6] Therefore it is not surprising that the probability $|\Psi(x, y, z_0)|^2$ to find the particle at some point (x,y) in the picture plane $z = z_0$, which follows from its solution, demonstrates the tell-tale interference pattern, and not just two blurry strips expected from a flow of classical particles (Figure 6.9).

This is a very remarkable result. Here too all attempts to explain away the wave-like behaviour of particles by assuming that it happens only when there are many of them, somehow forming a wave, were futile. Experiments with single electrons and other particles sent one by one towards a double slit (or a more elaborate structure, like a number of parallel slits – so-called diffraction grating) demonstrated the same result: an interference pattern. In the two-slit case this can only mean one thing: a particle passes through *both* slits. In case of a diffraction grating it passes through *all* the slits at once. Nevertheless it is registered as a particle by a detector (e.g. a sensitive film) in the picture plane. Recently such behaviour was observed for a molecule containing almost a thousand atoms, weighing about 10,000 atomic mass units (that is, about 10,000 times heavier than a proton or a neutron, and about 20 million times heavier than an electron).

In its time this experimental result produced a kind of shock, despite the fact that Louis de Broglie predicted the outcome three years earlier. The 'particle-wave duality' became a stock phrase, with some unfortunate consequences. Not for physics or physicists – physicists know what that means (that quantum objects are neither particles nor waves, though in certain

---

6   We will consider another, very elegant and illuminating way of finding the solution using *path integrals* in Chapter 9.

conditions they can behave as either). But this expression has the potential of thoroughly confusing non-physicists, and sometimes even creating in them an impression that quantum physics is self-contradictory (which it is not) or impossible to learn (which it is not either – though a student needs some love or at least tolerance of mathematics. But if you have managed to read this book this far, you should be fine.)

The de Broglie hypothesis was formulated when there was yet neither Heisenberg nor Schrödinger equations (it actually helped Schrödinger to formulate his version of quantum mechanics). It stated that any quantum object with momentum p should be accompanied by some wave with the wavelength h/p *(de Broglie wavelength)*.

Spotlight: Thomson, Davisson and Germer

George Thomson, Clinton Davisson and Lester Germer observed electron diffraction on a crystal in 1927. Thomas and Davisson were awarded the Nobel Prize in Physics in 1937.

Spotlight: Louis-Victor-Pierre-Raymond, Duke de Broglie

Louis-Victor-Pierre-Raymond, 7th Duke de Broglie (1892–1987) was not a duke (he inherited the title from his brother later) when he formulated his famous hypothesis, which was rewarded with the Nobel Prize for Physics in 1929 (after it was confirmed by the electron diffraction experiments).

Looking back to Chapter 3, you will see that we have already derived this very relation (as $p = \hbar k = h / \lambda$) for a plane wave solution of the Schrödinger equation. But de Broglie guessed it, based on some general considerations (including special relativity) and the Born–Sommerfeld quantization condition. The latter, with its requirement $\oint p \, dq = nh$, indeed brought to mind the resonance condition that an integer number of waves fit certain length, in our case the length of a closed particle trajectory:

$$L = n\lambda = \frac{nh}{p}.$$

But it took a great deal of vision and intellectual bravery to come to this conclusion and make it public. Science is not a game for the timid.

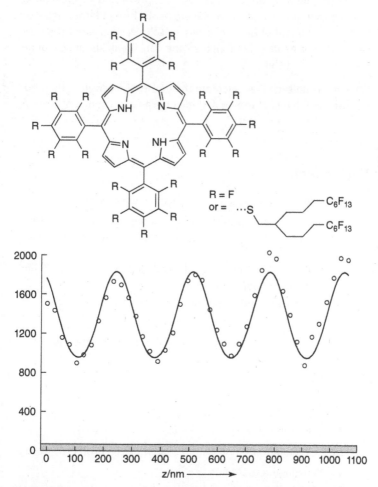

Figure 6.9 Here the quantum interference is demonstrated with *heavy* molecules: their masses exceed 10,000 amu (approximately 10,000 masses of a hydrogen atom) (Eiberberger et al., 2013, 14696–14700, Figs 1 and 3)

Case study: De Broglie wavelength

Consider the De Broglie wavelength of a particle with momentum p

$\lambda = h/p$.

For the heavy molecule in the experiment depicted in Figure 6.9, the de Broglie wavelength was 500 fm, approximately 10,000 times smaller than the size of the molecule itself. 1 fm (a femtometre or a 'fermi' in honour of Enrico Fermi) is 10^{-15}m and is about the size of a proton or neutron (as much as one can speak about size of an elementary particle).

You can, as an exercise, calculate the de Broglie wavelength of our ant (weight 0.5 mg) climbing a blade of grass at a speed of 1 mm/s.

Fact-check

1 Tunnelling was proposed as the mechanism of α-decay, because

   a the energies of α particles do not depend on temperature

   b the energies of α particles coming from the same decay are identical

   c the energy of α particles strongly depends on the half-life of the α-radioactive element

   d radium-222 and polonium-210 have almost identical half-lives

2 The probability to penetrate a potential barrier depends on its thickness

   a linearly

   b quadratically

   c exponentially

   d logarithmically

3 The scanning tunnelling microscope must be protected from

   a thermal fluctuations

   b cosmic rays

   c vibrations

   d electromagnetic noise

4 The subatomic STM resolution is due to

   a   the subatomic scanning tip

   b   the exponential dependence of tunnelling probability on distance

   c   the shortwave probing radiation

   d   electrons being smaller than the nuclei

5 The STM image shows

   a   electron density

   b   proton density

   c   current density

   d   light intensity

6 The lateral resolution of the STM is 0.1 nm. Therefore the uncertainty of the lateral momentum of the tip cannot be less than approximately

   a 10^{-12} kg . m

   b 10^{-23} kg . m

   c 10^{-24} kg . m

   d 10^{-43} kg . m

7 In photoelectric effect the 'red threshold' frequency is determined by

   a   the external potential

   b   the light intensity

   c   the light pressure

   d   the work function

8 The double slit experiment with single photons produces interference pattern because

   a   a photon is a wave

   b   a photon passes through both slits at once

   c   many single photons form a wave

   d   a photon is a massless particle

9 Double slit interference with a tennis ball will fail because

   a   it does not have a wave function

   b   its de Broglie wavelength is too short

   c   quantum mechanics does not apply

   d   tennis balls are elastic

10 The de Broglie wavelength (in metres) of an ant from the problem is

    a 1.32×10^{-8}

    b 1.32×10^{-15}

    c 1.32×10^{-24}

    d 1.32×10^{-32}

Dig deeper

R. Feynman, R. Leighton and M. Sands, *Feynman Lectures on Physics*. Basic Books, 2010 (there are earlier editions). Vol. I, Chapter 37 and Vol. III, Chapters 1–3.

E. H. Wichmann, *Quantum Physics* (Berkeley Physics Course, Vol. 4), McGraw-Hill College, 1971. Chapter 7.

7

The game of numbers

The general formalism of quantum mechanics, which we have described so far, does not care about the number of particles in the quantum system. The quantum state $|\Psi\rangle$ satisfies the same Schrödinger equation, $i\hbar\frac{\partial|\Psi\rangle}{\partial t} = \hat{H}|\Psi\rangle$, whether it describes a single electron, an electron beam, an atom, a molecule or a qubit (Figure 7.1). Similarly, the expectation value of an observable $\hat{A}$ is given by $\langle\Psi|\hat{A}|\Psi\rangle$, and the Heisenberg equation for it is always $i\hbar\frac{dA}{dt} = [\hat{A}, \hat{H}]$. The devil is, as usual, in the details. We know – at least, nominally – how to find the wave function $\Psi(x, y, z, t)$, which represents the state vector of a single quantum particle (e.g. an electron). Using it for an atom or a molecule (like we did describing certain quantum interference and diffraction experiments) is already a stretch – an approximation, which only holds while we can neglect both the size of the object so described and its interaction with other such objects (and, however strange that sounds, the latter can never be neglected – we shall see why later in this chapter). Therefore single-particle quantum mechanics only takes us so far. It needs further elaboration.

Figure 7.1 Hydrogen atom

Electron in an atom

Before qubits, the simplest quantum system used as a textbook example was the hydrogen atom. The atom consists of the nucleus (a single proton with the electric charge +1 in electron charge units) and an electron. The proton is almost 2000 times more massive than the electron[1], and therefore can be, to a good accuracy, considered as 'nailed down' – being at rest at the origin. Then its only role (neglecting some subtler effects we are not ready to treat yet) is to be the source of the electrostatic Coulomb potential,

$$\Phi(r) = +\frac{|e|}{r},$$

acting on the electron.[2] The stationary Schrödinger equation for the electron with energy E is thus

$$-\frac{\hbar^2}{2m_e}\nabla^2\Psi(R) - \frac{e^2}{r}\Psi(R) = E\Psi(R).$$

---

1   $m_p \approx 1836\, m_e$. If we want to be more accurate, we can replace the electron mass in the equation with the reduced mass,

$\mu = m_e \dfrac{m_p}{m_p + m_e} \approx m_e \left(1 - \dfrac{m_e}{m_p}\right)$. This will produce a correction of order 1/1836.

2   We use here the so-called Gaussian units for electromagnetic quantities. The international (SI) units (coulombs for charge, amperes for current, volts for voltage, etc.) are practical, but formulas written for them are quite unnecessarily ugly (e.g. the Coulomb potential would look like $\Phi(r) = +\dfrac{|e|}{4\pi\epsilon_0 r}$) . Theoretical physicists use Gaussian units and translate the results to SI in the end, if at all.

Here **R** is the radius vector pointing at the electron; its length is r. The symbol ∇^2 is pronounced *nabla square* and is called the *Laplace operator*, or simply the *Laplacian*. This is shorthand for the expression we used before: for any function of Cartesian coordinates *(x, y, z)*:

$$\nabla^2 f(x,y,z) = \left(\frac{\partial^2}{\partial x^2} + \frac{\partial^2}{\partial y^2} + \frac{\partial^2}{\partial z^2} \right) f(x,y,z).$$

Key idea: Laplacian operator and Schrödinger equations in spherical coordinates

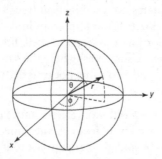

Cartesian coordinates are not always convenient, for example, when the system has what is called *spherical symmetry*, like the hydrogen atom. Indeed, the electrostatic potential only depends on the distance from the proton – in other words, the potential and therefore the Schrödinger equation has the same symmetry as an ideal sphere. (This does *not* mean that the solutions of the Schrödinger equation will have the same symmetry!)

In spherical coordinates the Schrödinger equation for an electron in the hydrogen atom looks as follows:

$$-\frac{\hbar^2}{2m_e} \frac{1}{r^2} \frac{\partial}{\partial r} \left(r^2 \frac{\partial}{\partial r} \Psi(r,\theta,\phi) \right) - \frac{\hbar^2}{2m_e} \frac{1}{r^2 \sin\theta} \frac{\partial}{\partial \theta} \left(\sin\theta \frac{\partial}{\partial \theta} \Psi(r,\theta,\phi) \right)$$
$$-\frac{\hbar^2}{2m_e} \frac{1}{r^2 \sin^2\theta} \frac{\partial^2}{\partial \phi^2} \Psi(r,\theta,\phi) - \frac{e^2}{r} \Psi(r,\theta,\phi) = E\Psi(r,\theta,\phi).$$

But in our case, when the potential depends only on the distance r between the electron and the proton, it is not convenient to use the Cartesian coordinates. The solution is greatly simplified if spherical coordinates are used instead (recall Chapter 4), and in these coordinates (r,θ,ϕ) the Laplacian looks quite differently. There are other coordinate systems as well, fitted to specific problem symmetries. Using everywhere the same symbol ∇^2 shortens equations and allows us not to specify a particular coordinate system.

We chose to write the Coulomb (electrostatic) energy of the proton as $-\frac{e^2}{r}$. (As we know from classical mechanics, any constant value can be added to energy without changing anything – it is the energy *differences* that matter.) Then at an infinitely large distance from the proton this potential energy is zero. If an electron manages to get to the infinity with some energy to spare, it actually breaks loose from the proton – there is now a free electron and a proton (or, if you wish, a positively charged *hydrogen ion*) instead of a hydrogen atom. Such an electron will have energy $E > 0$. Therefore as far as we are interested in a hydrogen atom, we must consider the solutions of the Schrödinger with $E \leq 0$ (the *bound states* of the electron).

Far away from the proton, when $r \to \infty$, we can drop from the Schrödinger equation all terms proportional to $1/r$ or $1/r^2$. What remains is

$$-\frac{\hbar^2}{2m_e}\frac{\partial^2}{\partial r^2}\Psi = E\,\Psi.$$

For a positive E the solution is $\Psi(r) = C\exp\left[\pm ir\sqrt{Em_e/\hbar^2}\right]$, while for a negative E it is

$$\Psi(r) = C \exp\left[\pm r \sqrt{|E| m_e / \hbar^2} \right], r \to \infty.$$

This determines what is called the *asymptotic* behaviour of the solution, its main feature at large distances from the proton. In the latter formula we must pick the minus sign in the exponent: an electron wave function, which grows exponentially at an infinite distance from the nucleus, is unphysical.

It turns out that while solutions exist for any positive E (which is reasonable – a free electron can have any energy it likes), only solutions with certain negative energies are allowed. The restrictions come from some reasonable requirements: that the wave function could be normalized (that is, we must be certain to find the electron *somewhere* in space); and that the wave function has a single value at any point. The allowed energy values can be all numbered – they form a *discrete set,* like the energy levels of a particle in the box featured in Chapter 3.

$$E_n = -\frac{\hbar^2}{2m_e a_0^2} \frac{1}{n^2}$$

The number $n = 1, 2,\ldots$ is called the *principal quantum number*, and $a_0 = \frac{\hbar^2}{m_e e^2}$ is the *Bohr radius* (which is a natural length scale for a hydrogen atom). You see that as the principal quantum number increases, the energy levels get higher and higher up and closer and closer to each other and to $E = 0$. The lowest energy state – the ground state of the electron in a hydrogen atom – is for $n = 1$. The corresponding energy scale, $-E_1 = \frac{\hbar^2}{2m_e a_0^2} \approx 13.6 \text{ eV}$, is called a *Rydberg* (and denoted by Ry). This is the ionization

energy – the minimal energy one must supply to the hydrogen atom in this state to break the electron loose and produce the positively charged hydrogen ion – that is, a proton.

The quantum formula for the electron energy levels explains the emission (and absorption) spectra of hydrogen (Figure 7.2). Suppose the hydrogen atom emits a photon. This can only happen if an electron drops from a level E_m to a level $E_n (m > n)$. The energy difference, carried away by the photon, is

$$\Delta E = h\upsilon = E_n - E_m = 1\text{Ry} \cdot \left(\frac{1}{n^2} - \frac{1}{m^2} \right).$$

If we rewrite it for the inverse wavelength of the emitted light,

$$\frac{1}{\lambda} = \frac{1\text{Ry}}{hc} \cdot \left(\frac{1}{n^2} - \frac{1}{m^2} \right) = R_H \left(\frac{1}{n^2} - \frac{1}{m^2} \right).$$

we obtain the *Rydberg formula*, which has been known since the 1880s but could not be explained within the framework of classical physics. (R_H is the *Rydberg constant*).

The energy of electron only depends on the quantum number n. But the wave functions also depend on the other two quantum numbers, l and m, which appear in the general formula for $\Psi(\mathbf{r}, \theta, \phi)$. These are called the *orbital* quantum number and the *magnetic* quantum number, and for any given n

$$0 \leq l \leq n-1; -l \leq m \leq l.$$

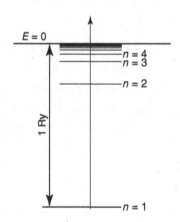

Figure 7.2 Energy levels and ionization energy in a hydrogen atom

Key idea: Bohr radius

Bohr radius $a_0 = \dfrac{\hbar^2}{m_e e^2} \approx 0.53$ Å (one angstrom is one hundred millionth part of a centimetre).

Spotlight: Johann Balmer and Johannes Rydberg

Johann Jakob Balmer (1825–1898) was a Swiss mathematician, who in 1885 found an empirical formula that quantitatively described the frequencies of visible light emitted by hydrogen. (Some say he did it on a bet – a friend dared him to find a simple formula fitting the hydrogen spectroscopic data.)

In 1888 Johannes Robert Rydberg (1854–1919), a Swedish physicist, found the general phenomenological formula, which in modern notation looks like:

$$\frac{1}{\lambda} = R_H \left(\frac{1}{n^2} - \frac{1}{m^2} \right).$$

Balmer's formula corresponds to the case $n = 2$.

That is, for each n there are n solutions with different l, and for each l there are $(2l+1)$ solutions with different m. The total number of independent solutions with the same energy En is thus $N_n = \sum_{l=0}^{n-1} (2l + 1) = n^2$.

The orbital quantum number determines the shape of the electron wave function (Figure 7.3). For historical reasons, some orbital states have also letter labels: s $(l = 0)$[3]; p $(l = 1)$; d $(l = 2)$; f $(l = 3)$; g $(l = 4)$; h $(l = 5)$; i $(l = 6)$.

Electrons in an atom and Mendeleev periodic table

The solutions obtained above work – with certain accuracy – also for heavier atoms with an atomic number Z. One should replace e^2 with Ze^2 (or a_0 with a_0/Z). Accuracy of this procedure is limited in the first place by the fact that electrons are

---

3 This wave function does not depend on angles and is therefore spherically symmetric; nevertheless 's' comes not from 'spherical', but from 'sharp'. Go figure.

interacting not only with the nucleus; they repel each other with Coulomb forces. There are other, weaker and subtler, interactions as well, but we can dismiss them for the moment. But the very fact that such a procedure works at all (though quite crudely) for atoms with several electrons is surprising.

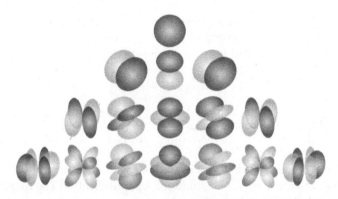

Figure 7.3 Shapes of s, p, d, f wave functions (the shade marks the sign of the wave function) (Sarxos 2007)

Let us reflect. Any physical system tends to its lowest energy state. If electrons would not repulse each other, they should all be in the state E_1. And even if we take their repulsion into account, they would try to occupy this state, which would make their interactions non-negligible and render the results obtained for a single electron totally inapplicable to the heavier than hydrogen elements ($Z > 1$).

For some reason this does not happen and instead electrons behave in a very orderly matter. In the ground state of an atom with $Z = 2$ (helium), the two electrons are both found in the ($n = 1, l = 0$) state. This *electronic structure* is usually denoted by $1s^2$.

For $Z = 3$ (lithium), there are two s-electrons with $n = 1$ and one s-electron with $n = 2$, that is, $1s^2 2s^1$. The next element, $Z = 4$ (beryllium), has the structure $1s^2 2s^2$. We see that a given set of quantum numbers can be 'shared' by two electrons, no more, and they are being taken starting from the lowest energy state – like a rooming house being occupied starting from the ground floor (Figure 7.4).

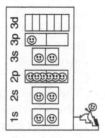

Figure 7.4 Electrons moving in

Table 7.1 Electronic structures for $Z = 5...10$

$Z = 1$	Hydrogen	$1s^1$
2	Helium	$1s^2$
3	Lithium	$1s^2 2s^1$
4	Beryllium	$1s^2 2s^2$
5	Boron	$1s^2 2s^2 2p^1$
6	Carbon	$1s^2 2s^2 2p^2$
7	Nitrogen	$1s^2 2s^2 2p^3$
8	Oxygen	$1s^2 2s^2 2p^4$
9	Fluoride	$1s^2 2s^2 2p^5$
10	Neon	$1s^2 2s^2 2p^6$

Recalling that for $n = 2$ the orbital quantum number can take values $l = 0$ and $l = 1$, and that there are three ($2 \cdot 1 + 1$) different-m states corresponding to $l = 1$ (p-states), we can write the electronic structures for $Z = 5...10$ (we have also included the previous results for completeness):

In this way we have filled in the first and second rows of Mendeleev's Periodic Table. We could continue, and – giving special consideration to the lanthanide and actinide elements – would faithfully reproduce all of it.

This exercise makes sense. The chemical properties of the elements are determined by the behaviour of their electrons, mostly the electrons with the highest energy (biggest n), which are the easiest to remove (recall, that $E_n \sim -1/n^2$). Even though due to electron–electron interactions and other corrections the states with the same n but different l will no longer have exactly the same energy (the technical term is *'the degeneracy is lifted'*), this observation remains valid. The dependence of chemical properties on the electronic configuration is not only due to the number of electrons in the states with the greatest n, but also due to the fact that different solutions of the Schrödinger equation have different shapes – that is, different probabilities $|\Psi(r,\theta,\phi)|^2 = |\Psi(r)|^2$ to find the electron at a given point r and, therefore, the possibility of directed couplings between atoms.[4] The repeated pattern of electron states explains the periodic dependence of the chemical properties on the atomic number Z. For example, all elements with full electron shells (that is, when every available state is occupied by exactly two electrons) are the noble gases (helium, neon, argon, krypton, xenon and radon), which due to this completeness practically do not participate in chemical reactions. All elements with a single s-electron in the outermost shell (hydrogen and the alkali metals – lithium, sodium, potassium, rubidium, caesium and francium) are, on the contrary, very active chemically.

---

4   Ancient atomists imagined that their 'atoms' – tiny, solid, indestructible particles – have various shapes with something like hooks and eyes, with which they connect to each other in order to form macroscopic bodies. What we discuss here is a vastly different concept, but it is still eerily reminiscent of the insights of these men of genius.

Remarkably the scheme generally holds even though the hydrogen-like formulas only work properly either for one-electron ions (when a single electron remains bound to the nucleus), or for the *Rydberg atoms*, when one electron occupies such a high energy level $n \gg 1$, that on average it is very far away both from the nucleus and from the rest of electrons, which keep close to the nucleus. Generally, 'hydrogen-like' electron wave functions are distorted by electron–electron interactions out of all recognition. For example, in a carbon atom, its four electrons with $n = 2$, instead of occupying a spherically symmetric s-state and dumbbell-like p-states along x, y and z-axes, occupy four identical states directed at the vertices of a tetrahedron (Figure 7.5). (This is very lucky for us – it is this electronic structure of carbon that makes it such a useful building block for all organic compounds, including those our brain is made of.) Nevertheless, the systematics of electron states based on the hydrogen-like solutions of a single-electron Schrödinger equation for some reason remains valid.

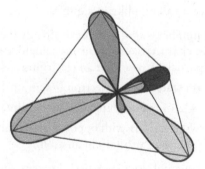

Figure 7.5 Carbon atom

Pauli exclusion principle, fermions and bosons, and spin

To begin with, let us return to the question why electrons occupy these states in such an orderly way. As a matter of fact, they have no choice. They obey the *Pauli exclusion principle*. This principle states that **two identical fermions cannot occupy the same quantum state but any number of identical bosons can occupy the same quantum state.**

Examples of *fermions* (named after Enrico Fermi) are electrons, protons, neutrons and some more exotic elementary particles. There are other quantum particles, which do not satisfy the exclusion principle (for example, photons). They are called *bosons*. It seems that whoever designed this Universe believed that whatever is not prohibited is mandatory, since the bosons follow the exact opposite rule: there can be any number of bosons in the same quantum state.

The Pauli principle helps understand why electron–electron repulsion does not completely mess up the quantum states in an atom – since only one electron is permitted per state, they would have to hold somewhat apart from each other, reducing the Coulomb repulsion. But since there are *two* electrons per state (otherwise our state counting would not reproduce the Mendeleev table), one must assume that there is something that distinguishes these electrons and makes it possible for them to have the same wave function. In other words, in addition to the quantum numbers *n, l, m* we must introduce one more, which can only take two values – like in a qubit.

This quantum number – *spin* – we have already mentioned in Chapter 4. Spin characterizes the intrinsic angular momentum of a quantum particle: $L = \hbar s$. For an electron $s = \frac{1}{2}$, this yields exactly two extra states for an electron – 'spin up' $\left(+\frac{1}{2}\right)$ and 'spin down' $\left(-\frac{1}{2}\right)$ (of course, 'up' and 'down' directions are arbitrary). So two electrons with opposite spins can share each electron wave function ('orbital') in an atom.

There is a fundamental 'spin-statistics' theorem, due to Pauli, which says that *all* particles with half-integer spins are fermions, and *all* particles with integer spins are bosons. 'Statistics' indicates that statistical properties of large groups of bosons and fermions are very different, as we shall see.

We cannot prove this theorem here, but we can prove another neat statement – that all quantum particles are *either* fermions *or* bosons. The proof uses only one assumption: that any two elementary particles of the same kind are impossible to distinguish.

Case study: Principle of indistinguishability of elementary particles

The principle of indistinguishability of elementary particles states that there is absolutely no way to tell apart any two particles of the same kind. You cannot do so even in a Gedankenexperiment (e.g. by painting one electron red). This principle is supported by experimental evidence. That is, if it were not satisfied, the observable properties of matter and radiation would be massively different.

Then the proof is straightforward. Consider a quantum state $|\Psi\rangle$ of a quantum system, which contains N identical particles. We can write this state as a wave function in the configuration space, that is, dependent on the coordinates (and spins!) of all particles:

$$|\Psi\rangle \to \Psi(r_1 s_1, r_2 s_2, \ldots r_N s_N).$$

Its square modulus $|\Psi(r_1 s_1, r_2 s_2, \ldots r_N s_N)|^2$ gives the probability density to find one particle with spin projection s_1 at the point r_1, and another with spin projection, s_2 at the point r_2, etc. Now let any two particles change places. Since they are impossible to distinguish, all the probabilities (i.e. the modulus of the wave function) must stay the same. Therefore the wave function can only be multiplied by some factor γ such that $|\gamma| = 1$:

$$\Psi \to \gamma \Psi.$$

Key idea: Quantum angular momentum $\hat{L}$

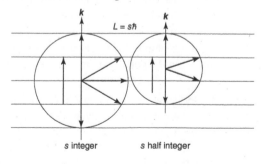

s integer   s half integer

Now let us exchange these two particles *again*. On the one hand, we must again multiply the wave function by the same factor γ. On the other hand, since both particles are back to their original places, the wave function must be *exactly the same* as before we started moving particles around. Therefore $\gamma^2 = 1$.

Spotlight: George Uhlenbeck and Samuel Goudsmit

Two very young theoretical physicists, George Uhlenbeck (1900–1988) and Samuel Goudsmit (1902–1978) discovered spin in 1925 (the latter still a graduate student). They proposed the idea of a 'spinning' electron, with quantized angular momentum $\frac{\hbar}{2}$, in order to explain some experimental data. As a well-known story of this discovery goes they attempted to withdraw their paper before publication, because a spinning electron (if you imagine it as a solid ball) would have points on its surface moving faster than light and thus violate the relativity theory. Fortunately, their boss was Paul Ehrenfest, who insisted on publication, since 'Well, that is a nice idea, though it may be wrong. But you don't yet have a reputation, so you have nothing to lose'. Ehrenfest was right, as we now know: an electron is *not* a solid ball, and spin is *not* due to its rotation.

Ralph Kronig (1904–1995) was not so lucky – his idea of the 'rotating electron' proposed independently early in 1925 was shot down by Pauli as having 'nothing to do with reality', and Kronig decided not to publish it.

This leaves us with two possibilities. Either $\gamma = 1$ (that is, the wave function does not change at all, if two identical particles are exchanged), or $\gamma = -1$ (the wave function changes sign). In the former case we have bosons – named after Satyendra Nath Bose, and in the latter – fermions.

A 'rule of thumb' allowing us to distinguish between bosons and fermions is, that fermions are 'matter' (or 'substance'), while bosons are 'fields'. For example, substances are made of atoms, atoms are made of electrons, protons and neutrons, and protons and neutrons are 'made of' quarks. All of these are fermions with $s = 1/2$. On the other hand, light quanta – photons – are bosons (with $s = 1$), and all electric and magnetic forces binding electric charges also can be described in terms of photons. The celebrated Higgs boson has spin $s = 0$. It is expected that the gravity quanta are bosons with $s = 2$. (They were not observed yet and a satisfactory quantum theory of gravity remains to be built.)

The distinction agrees with our intuition that 'substances' occupy space, and you cannot put one 'substance' into the space already occupied by another.

Key idea: 'Hardcore' compound bosons

An important notice: fermions can 'lump together' to form compound bosons. For example, α particles are bosons, even though they consist of four fermions (two protons and two neutrons), because their spins point in the opposite directions, and the overall spin is zero. The helium-4 atoms (an α particle and two s-electrons) are bosons as well. On the other hand, bosons cannot form fermions, no matter what.

Before going on, it is worth mentioning that physicists – especially condensed matter physicists – quite often talk about and experiment with particles, which are neither bosons nor fermions. These are not 'real' particles in three-dimensional space. For example, if you confine electrons in a semiconductor to a single plane or even a single line (which actually can be done), they can behave *as if* they were such hybrid particles. This does not invalidate our theorem. In a two- or one-dimensional system one cannot make two particles change places quite as easily, as in three (or more) dimensions (try switching places on a packed bus!), and this makes the difference.

Quantum many-body systems, Fock states, Fock space and creation and annihilation operators

We have seen that the quantum state of a system of N identical quantum particles must be either symmetric (i.e. remain the same) or antisymmetric (i.e. change sign) with respect to exchange of any two particles. Taking this into account explicitly is time-consuming and boring work, but in many cases it can be avoided altogether.

In order to do so we need a convenient Hilbert space. Let us assume that we know *one-particle* states for a system of N free, non-interacting particles. Such states can be, for example, plane waves. As we already know from Chapter 3, a one-dimensional plane wave, exp[ikr], describes a particle with a definite momentum $\hbar k$. In three dimensions these expressions, obviously, will turn into exp[$i k.r$] and $\hbar k$. Or we can take instead states, where a particle has a definite position, r.

Now consider a quantum state, which has n_1 particles in state *1*, n_2 particles in state *2*, n_3 in state *3* etc., and *has the appropriate symmetry* – that is, is symmetric for bosons and antisymmetric for fermions. We denote it by $|n_1, n_2, n_3, ...\rangle$. Numbers n_1, n_2, n_3 are called the *occupation numbers*. Then let us introduce two special quantum operators – a *creation* and an *annihilation* operator. They are denoted by c_i^+ and c_i and, according to their names, create or annihilate particles in a given (i^{th}) one-particle state:

$$c_i^+ |n_1, n_2, ... n_i, ...\rangle \propto |n_1, n_2, ... n_i + 1, ...\rangle;$$
$$c_i |n_1, n_2, ... n_i, ...\rangle \propto |n_1, n_2, ... n_i - 1, ...\rangle.$$

The factors in these expressions are different for bosons and for fermions.

First, note that creation and annihilation operators c_j^+ and c_j do not correspond to any observables. They are not Hermitian – as a matter of fact, they are Hermitian conjugates of each other (we have seen examples of this in Chapter 4). But their combinations may be Hermitian.

Key idea: Hermitian or not?

You can check that both $(c_j^+ + c_j)$ and $i(c_j^+ - c_j)$ are Hermitian (here i is the imaginary unit). For example,

$$\left[i\left(c_j^+ - c_j\right)\right]^+ = -i\left(\left(c_j^+\right)^+ - \left(c_j\right)^+\right) = i\left(c_j^+ - c_j\right).$$

(Recall, that $(c_j)^+ = c_j^+, (c_j^+)^+ = c_j$, and that when taking a Hermitian conjugate we also replace complex numbers with their conjugate: $i \to i^* = -i$.

Second, these operators are convenient if we want to describe any changes in the system. Suppose a particle changed its state (e.g. momentum or position) from one labelled by 'j' to one labelled by 'k'. Since particles cannot be distinguished as a matter of principle, this means that now there is one particle less in state 'j' and one more in state 'k'. Such a change can be affected by acting on the many-particle state first with c_j and then with c_k^+:

$$c_k^+ c_j \left|\ldots n_j, \ldots n_k, \ldots\right\rangle \propto \left|\ldots n_j - 1, \ldots n_k + 1, \ldots\right\rangle.$$

Third, generally, these operators do not commute. Moreover, it turns out that their commutation relations are of the paramount importance. If creation and annihilation operators *commute*, that is,

$$\left[c_j, c_k^+\right] = c_j c_k^+ - c_k^+ c_j = \delta_{jk}, \text{ and} \left[c_j, c_k\right] = \left[c_j^+, c_k^+\right] = 0,$$

they describe bosons and are called *Bose operators*. If they *anticommute*, that is,

$$\left\{c_j, c_k^+\right\} = c_j c_k^+ + c_k^+ c_j = \delta_{jk}, \text{ and} \left\{c_j, c_k\right\} = \left\{c_j^+, c_k^+\right\} = 0,$$

they describe fermions and are called *Fermi operators*. Note that two creation or two annihilation operators simply commute (anticommute), while an annihilation and a creation operator (for the same single-particle state) commute (anticommute) to the unity.

Key idea: Kronecker delta

The symbol δ_{jk} is called the Kronecker delta (after a German mathematician). It equals unity, if $j = k$, and zero otherwise.

Fourth, the quantum state, which does not contain any particles, is called the *vacuum state* and is denoted by $|0\rangle$:

$$|0\rangle = |0,0,0,\ldots0,\ldots\rangle.$$

This is, of course, the lowest energy state – the ground state – of the system. Since there are no particles to annihilate, if you act on the vacuum state by any annihilation operator, you obtain zero:

$$c_j |0\rangle = 0.$$

Note that the vacuum state is *not* zero – a vacuum is not 'nothing'! On the contrary, any physical state of the system can be obtained by acting on the vacuum state by the appropriate creation operators as many times as necessary (the nth power of an operator means that the operator was acting n times in a row):

$$\left|\ldots n_j,\cdots n_k,\cdots\right\rangle \propto \left(c_j^+\right)^{n_j}\left(c_k^+\right)^{n_k}\left|\ldots n_j,\cdots n_k,\cdots\right\rangle.$$

The vacuum state and all states that can be obtained from it in this way are called the *Fock states*. They form the basis of a special kind of Hilbert space. It is called the *Fock space* (after Vladimir Aleksandrovich Fock).

Let us see how this works. We will start with fermions.

The anticommutation relation for two Fermi creation operators means that we cannot create more than one particle in a given state (say, state j). Indeed, in order to create a quantum state with two fermions in a state j, one must twice act on the vacuum state by the creation operator c_j^+. Since $\left\{c_j^+, c_j^+\right\} = 2c_j^+ c_j^+ = 2\left(c_j^+\right)^2 = 0$, we will obtain zero as a result of this operation. Therefore the only possible states for a many-fermion system are of the form

$$|...0,1,1,0,1,...\rangle.$$

This is exactly what the Pauli principle requires: one fermion per state.

Of course, such a formula is only meaningful if you know *what* are the single-particle states, the occupation numbers of which are listed. Often the particle states with a given momentum and spin projection are used.

Using the Fock space is a very convenient way of dealing with a Hilbert space of a system of many (*arbitrary* many) identical particles – if you wish, like a convenient choice of coordinate system in geometry.

Speaking of coordinate systems, the basis of the Fock space of a system of fermions is naturally ordered:

$|0,0,0,0...\rangle, |1,0,0,0...\rangle, |0,1,0,0...\rangle, |1,1,0,0...\rangle, |0,0,1,0...\rangle$, etc.[5]

The operator

$$\hat{N}_j = c_j^+ c_j$$

is called the *particle number operator*. It turns out that it acts on a state $|\Psi\rangle = |...n_j,...n_k,...\rangle$ as follows:

$$c_j^+ c_j \left|...n_j,...n_k,...\right\rangle = n_j \left|...n_j,...n_k,...\right\rangle,$$

that is, it does not change the vector $|\Psi\rangle$, but simply multiplies it by the number of particles n_j in the single-particle state j (for example, by the number of particles with a given momentum and spin). As you recall, this is another way of saying that

---

5   You may have noticed, that these are natural numbers *0, 1, 2, 3, 4...* written *backwards* in the binary code.

the vector $|\Psi\rangle$ is an eigenstate of the operator $\hat{N}_j$ with the eigenvalue n_j.

Spotlight: Hermitian conjugates

A Hermitian conjugate of a product of operators (that is, a series of operators acting one after another) is given by

$$(\hat{A}\hat{B}\hat{C}...\hat{G})^+ = \hat{G}^+...\hat{C}^+\hat{B}^+\hat{A}^+.$$

Note the reversal of the order! You can check this for two operators presented by two-by-two matrices like those in Chapter 4.

For example, we can use as the single-particle states the hydrogen-like solutions for a single-electron Schrödinger equation:

'1'	$n = 1, l = 0, m = 0$, spin $-1/2$
'2'	$n = 1, l = 0, m = 0$, spin $+1/2$
'3'	$n = 2, l = 0, m = 0$, spin $-1/2$
'4'	$n = 2, l = 0, m = 0$, spin $+1/2$
'5'	$n = 2, l = 1, m = -1$, spin $-1/2$
'6'	$n = 2, l = 1, m = -1$, spin $+1/2$

and so on. Then $|1,1,0,...\rangle$ is the state $1s^2$.

The operator $\hat{N}_j$ is Hermitian and corresponds to an observable. Its expectation value is, predictably, n_j, since

$$\langle \hat{N}_j \rangle = \langle \Psi | \hat{N}_j | \Psi \rangle = \langle \Psi | n_j | \Psi \rangle = n_j \langle \Psi | \Psi \rangle = n_j.$$

Here we are talking about fermions, so n_j is always either zero or one.

In case of bosons, the only (but important) difference is that the operators commute, rather than anti-commute; the occupation numbers can be any non-negative integers; and so are the eigenvalues and expectation values of the particle number operator for any single-particle state j.

Finally, note that the expectation value of any c_j or c_j^+ in any of the Fock basis states is exactly zero:

$$\langle c_j \rangle = \langle n_1, n_2,... | c_j | n_1, n_2,... \rangle = 0; \langle c_j^+ \rangle = \langle n_1, n_2,... | c_j^+ | n_1, n_2,... \rangle = 0$$

This is reasonable: either of these operators changes the number by one in the state on the right, but not on the left; and the scalar product of two different basis vectors is zero.

Quantum oscillator, Heisenberg equations and energy quanta*

Creation and annihilation operators are indispensable when dealing with systems containing many quantum particles in different states, but they can be very useful in simpler cases as well. Consider, for example, the quantum harmonic oscillator, which we encountered in Chapter 2. Its classical counterpart in Chapter 1 has the Hamilton function (that is, energy expressed in terms of momentum rather than velocity):

$$H(x,p_x) = \frac{1}{2}m\omega_0^2 x^2 + \frac{1}{2}\frac{p_x^2}{m}.$$

From here we can obtain the Hamiltonian (operator!) for a quantum oscillator by replacing classical coordinates x and px with operators $\hat{x} = x$ and $\hat{p}_x = \frac{h}{i}\frac{d}{dx}$:

$$\hat{H} = \frac{1}{2}m\omega_0^2\hat{x}^2 + \frac{1}{2}\frac{\hat{p}_x^2}{m} = \frac{1}{2}m\omega_0^2 x^2 - \frac{\hbar^2}{2m}\frac{d^2}{dx^2}.$$

In order to find the quantized energy levels of this oscillator we would have to solve the Schrödinger equation,

$$-\frac{\hbar^2}{2m}\frac{d^2}{dx^2}\Psi(x) + \frac{1}{2}m\omega_0^2 x^2\Psi(x) = E\Psi(x).$$

We would find that the solutions are expressed in terms of some known functions, that they are physically meaningful (e.g. drop to zero for $|x| \to \infty$) only for certain values of energy, and in this straightforward but tiresome way we would eventually arrive at $E_n = \hbar\omega_0\left(n + \frac{1}{2}\right)$. But we are going to do nothing of the kind.

Instead we note that $[\hat{x}, \hat{p}_x] = i\hbar$ and that $[a, a^+] = 1$ for some Bose operators a, a^+. Let us introduce operators $\hat{X} = \frac{K}{\sqrt{2}}(a + a^+)$ and $\hat{P} = \frac{\hbar}{i\sqrt{2}K}(a - a^+)$. These operators are Hermitian and their commutator is

$$[\hat{X}, \hat{P}] = \frac{\hbar}{2i}[(a + a^+), (a - a^+)] = \frac{\hbar}{2i}(-[a, a^+] + [a^+, a]) = -\frac{\hbar}{i} = i\hbar.$$

We can therefore identify these operators with the original $\hat{x}$ and $\hat{p}_x$. But to what advantage? We will see it as soon as we substitute them in the Hamiltonian:

$$\hat{H} = \frac{1}{2}m\omega_0^2 \cdot \frac{K^2}{2}(a + a^+)^2 + \frac{1}{2m}\left(\frac{\hbar}{i\sqrt{2}K}\right)^2 (a - a^+)^2$$

$$= \frac{1}{4}m\omega_0^2 K^2(a^2 + aa^+ + a^+a + (a^+)^2) - \frac{\hbar^2}{4mK^2}(a^2 - aa^+ - a^+a + (a^+)^2).$$

(When taking the square of $(a \pm a^+)$ we did remember that aa^+ is not equal to a^+a.)

Now we notice that by a proper choice of the constant K we can eliminate from this expression all 'off-diagonal' terms (this is what the terms a^2 and $(a^+)^2$ are called):

$$\hat{H} = \frac{1}{4}(a^2 + (a^+)^2)\left\{m\omega_0^2 K^2 - \frac{\hbar^2}{mK^2}\right\} + \text{the rest}.$$

If choose $K^2 = \frac{\hbar}{m\omega_0}$, the expression in the curly brackets becomes zero, and the Hamiltonian will contain only 'the rest':

$$\hat{H} = \frac{1}{4}(aa^+ + a^+a)\left\{m\omega_0^2 K^2 + \frac{\hbar}{mK^2}\right\} = \frac{1}{4}(aa^+ + a^+a) \cdot 2\hbar\omega_0 = \hbar\omega_0 \frac{aa^+ + a^+a}{2}.$$

We are almost there. What remains is to recall that $aa^+ = a^+a + 1$ (which is just another way of writing the commutation relation). Therefore the Hamiltonian of a quantum harmonic oscillator is

$$\hat{H} = \hbar\omega_0\left(a^+a + \frac{1}{2}\right).$$

This Hamiltonian is expressed in terms of creation and annihilation operators. They act in the Fock space, which has the basis $|0\rangle, |1\rangle, |2\rangle\ldots$ (We have only one kind of operator, without any indices, so we need only one number to label a quantum state.)

Now we can easily find the quantized energy levels of harmonic oscillator. Remember, that the operator $\hat{N} = a^+ a$ is the particle number operator, and its expectation value in state $|n\rangle$ is simply n. Therefore the expectation value of the Hamiltonian in state $|n\rangle$ is

$$E_n = \left\langle n \left| \hbar\omega_0 \left(\hat{N} + \frac{1}{2} \right) \right| n \right\rangle = \hbar\omega_0 \left(n + \frac{1}{2} \right) = h\nu_0 \left(n + \frac{1}{2} \right).$$

We have obtained the Planck formula of Chapter 2, together with the zero-energy term $\frac{h\nu_0}{2}$! And all it took was some algebra (admittedly, algebra with operators, but still algebra).

You can easily check that a state $|n\rangle$ is actually an eigenstate of the Hamiltonian with energy E_n. This helps clarify what exactly *are* the particles, which our creation and annihilation operators a^+, a create and annihilate. They are the *energy quanta* of our oscillator. (And, as we have seen, they are bosons.)

We can do more. For example, we can find out what the time dependence of position and momentum of our oscillator will be. Since both are expressed through the operators a, a^+, it will be enough to find their dependence on time.

In order to do so, it is best to use the Heisenberg equations of motion:

$$i\hbar \frac{d}{dt} a = [a, \hat{H}] = \hbar\omega_0 [a, a^+ a] = \hbar\omega_0 a; \; i\hbar \frac{d}{dt} a^+ = \hbar\omega_0 [a^+, a^+ a] = -\hbar\omega_0 a.$$

The solutions of these equations are, as you can easily check, $a(t) = a(0) \exp(-i\omega_0 t)$ and $a^+(t) = a^+(0) \exp(i\omega_0 t)$. This immediately yields the expressions for $\hat{x}(t)$ and $\hat{p}_x(t)$.

When calculating the expectation value of the position or momentum using these expressions, we will be surprised – both are strictly zero at any time! (Which is not really surprising – the expectation value of either creation or annihilation operator in any state $|n\rangle$ is zero, as we have seen.)

This does not mean that each time we find the oscillator at the origin and at rest – just that it is equally likely to find it right or left of the origin, moving in either direction, so the average is zero. The expectation values of *squares* of position or momentum are clearly nonzero, since both are non-negative and

$$\langle n|\hat{H}|n\rangle = \frac{m\omega_0^2}{2}\langle n|\hat{x}^2|n\rangle + \frac{1}{2m}\langle n|\hat{p}_x^2|n\rangle = E_n > 0.$$

Quantum fields and second quantization

We have said that the approach based on creation and annihilation operators is custom-made for the description of many-particle systems. We saw how it works with electrons in an atom. But will it work with free electrons as conveniently? Will it work with photons?

Let us first consider a system of particles described by a Fock state vector $|\Psi\rangle = |n_1,n_2,n_3,\ldots n_j,\ldots\rangle$. We cannot actually say what this vector is unless we specify the single-particle states labelled by *1, 2, 3*, etc. For example, we can take as a state number *j* a state with wave function (one-particle wave function!) $\psi_j(r)$. These functions with different labels should be orthogonal (in the sense of the Hilbert space of functions, that is, the integral $\int d^3r\, \psi_j^*(r)\psi_l(r)$ is zero unless $j = l$).

An example of such set of functions is a set of plane waves,

$\psi_k(r) = C\exp(ik\cdot r)$ (*C* is a number – a normalization factor.) A particle in a state $\psi_k(r)$ has an exactly known momentum of $\hbar k$ and an indefinite position (because the probability of finding a particle $\propto |\psi_k(r)|^2 = |C|^2$ is the same at any point of space).

Now let us put together *field operators*:

$$\hat{\psi}(r) = \sum_j \psi_j(r)\, c_j; \hat{\psi}^+(r) = \sum_j \psi_j^*(r)\, c_j^+.$$

The operator $\hat{\psi}^+(r)$ creates a particle at the point r, and the operator $\hat{\psi}(r)$ annihilates it. The density of particles at this point is given by the operator $\hat{\psi}^+(r)\,\hat{\psi}(r)$. These operators describe the *particle field*, i.e. the situation when a particle can be found anywhere in space.

Spotlight: Particle density operator

The operator $\hat{n}(r) = \hat{\psi}^+(r)\,\hat{\psi}(r)$ describes the density of particles at a given point. Indeed, its expectation value is:

$$\langle\Psi|\hat{n}(r)|\Psi\rangle = \left\langle\Psi\left|\left(\sum_j\psi_j^*(r)c_j^+\right)\left(\sum_l\psi_l(r)c_l\right)\right|\Psi\right\rangle$$

$$= \sum_j\sum_l\psi_j^*(r)\psi_l(r)\left\langle\Psi\left|c_j^+c_l\right|\Psi\right\rangle$$

$$= \sum_{j=l}\left|\psi_j(r)\right|^2\left\langle\Psi\left|c_j^+c_j\right|\Psi\right\rangle + \sum_{j\neq l}\psi_j^*(r)\psi_l(r)\left\langle\Psi\left|c_j^+c_l\right|\Psi\right\rangle.$$

Each term in the second sum is zero. To see it, consider $\left\langle\Psi\left|c_j^+c_l\right|\Psi\right\rangle$ as a scalar product of two state vectors, $|\phi_l\rangle = c_l|\Psi\rangle$ and $|\phi_j\rangle = c_j|\Psi\rangle$. They are different ($|\phi_l\rangle$ has (n_l-1) particles in state l and n_j particles in state j, and $|\phi_j\rangle$ the other way around), and therefore their scalar product is zero.

In the first sum, $\left\langle\Psi\left|c_j^+c_j\right|\Psi\right\rangle = n_j$, the number of particles in state j. Therefore

$$\langle\Psi|\hat{n}(r)|\Psi\rangle = \sum_j\left|\psi_j(r)\right|^2 n_j$$

indeed gives the density of particles at the point r. The square modulus $\left|\psi_j(r)\right|^2$, with a correct normalization factor, is the local density of the 'one-particle probability cloud' described by the wave function $\psi_j(r)$, and n_j tells how many particles share this cloud.

Any observable can be expressed in terms of these field operators, similarly to how we expressed the Hamiltonian of a quantum oscillator in terms of the operators a, a^+. This is an almost automatic procedure.

Spotlight: Operators of potential and kinetic energy

The expectation value of the potential energy of an electron with the wave function $\psi(r)$ in an external electric potential $\phi(r)$ is given by $\langle\psi|U|\psi\rangle = \int d^3r\phi(r)[-|e|n(r)]$, where $n(r) = \psi^*(r)\psi(r)$ is the probability density of finding this electron at the point r. Then the potential energy *operator* for a many-electron system is

$$\hat{U} = -|e|\int d^3r\phi(r)\hat{\psi}^+(r)\hat{\psi}(r).$$

For the kinetic energy we obtain from

$\left\langle\psi\left|-\dfrac{\hbar^2}{2m}\nabla^2\right|\psi\right\rangle = \int d^3r\left(-\dfrac{\hbar^2}{2m}\right)\psi^*(r)\nabla^2\psi(r)$ the expression for the kinetic energy operator:

$$\hat{K} = \left(-\frac{\hbar^2}{2m}\right)\int d^3r\,\hat{\psi}^+(r)\nabla^2\hat{\psi}(r).$$

Now look what we have done. We have replaced a single-particle wave function $\psi(r)$ (which is a quantum state *vector* in a *single*-particle Hilbert space) by an *operator* $\hat{\psi}(r)$, which acts on quantum state vectors in the Fock space, i.e. in the Hilbert space of a *many*-particle system. Because of a superficial similarity of this operation to the transition from classical to quantum mechanics (when x, p_x, etc, were replaced with operators), the approach using creation and annihilation operators and Fock states is called the *second quantization*.

The strength of this approach is its ability to treat a system with an unlimited number of quantum particles and *automatically* take into account their statistics (whether they are bosons or fermions). As a matter of fact, it becomes indispensable even if you want to consider a single quantum particle (e.g. a single electron) accurately enough. And, of course, it is extremely handy when dealing with condensed matter systems, such as metals or semiconductors, which contain huge numbers of quantum particles. These questions we will consider in the next chapter.

Why classical light is a wave and a piece of metal is not

Before moving on, let us make a useful remark. One can always take a creation and an annihilation field operator and build out of them two *Hermitian* field operators, which is the way we obtained the position and momentum operators for a harmonic oscillator:

$$\frac{\hat{\psi}^+(r,t) + \hat{\psi}(r,t)}{\sqrt{2}} = \hat{\varphi}(r,t) = \hat{\varphi}^+(r,t);$$

$$\frac{\hat{\psi}^+(r,t) - \hat{\psi}(r,t)}{i\sqrt{2}} = \hat{\chi}(r,t) = \hat{\chi}^+(r,t)$$

Such operators do not necessarily have a physical meaning, but in case of bosons they usually do. For example, photons (light quanta) and phonons (sound quanta) are conveniently described by precisely such Hermitian operators.

There is an essential difference between the commutation relations of Fermi and Bose operators. For bosons, the *commutator* is proportional to the unity. If we consider the classical situation, when there are *very* many bosons, the unity can be neglected compared to their number. Therefore the Hermitian field operators could be treated just as normal numbers, like directly measurable field amplitudes. (They must be Hermitian, since we cannot measure complex numbers). So in a classical light wave $\hat{\varphi}(r,t), \hat{\chi}(r,t)$ become the amplitudes of the electric field with given polarizations, and light remains a wave even in the classical limit.

For fermions the situation is drastically different: their field operators *anticommute*. You can check that for fermions

$$\{\hat{\varphi},\hat{\varphi}\} = 0; \{\hat{\chi},\hat{\chi}\} = 0; \{\hat{\varphi},\hat{\chi}\} = 0.$$

Therefore no matter what, the fermion wave functions cannot become usual waves in the classical limit: such a 'wave' would have to have zero amplitude. So, e.g. a peace of metal, which consists of fermions, must be solid to exist.[6]

 Fact-check

1 The principal quantum number determines

   a the bound state energy of an electron
   b the ionization energy
   c the number of electrons in an atom
   d the Rydberg constant

2 The orbital quantum numbers determine

   a the momentum of an electron in an atom
   b the shape of the electron wave function
   c the number of electrons in an atom
   d the energy of an electron in an atom

3 The Pauli exclusion principle states that

   a two quantum particles cannot occupy the same quantum state
   b two fermions cannot occupy the same quantum state
   c no more than two bosons can occupy the same quantum state
   d electrons have spins

---

6 You may ask: 'What about atoms, which are bosons – like helium-4?' The answer is, compound bosons are what is called 'hard core bosons' – in the sense that they behave like bosons only until you try pushing them too close together.

4 Spin characterizes

    a   an orbital motion of a quantum particle

    b   an orbital angular momentum of a quantum particle

    c   an intrinsic angular moment of a quantum particle

    d   the number of electrons in an atom

5 If a particle has spin angular momentum $L = \frac{7}{2}\hbar$, then

    a   the projection of its angular momentum on z-axis may take seven values

    b   may take eight values

    c   this particle is a boson

    d   this particle is a fermion

6 If two particles are exchanged, the wave function of a system of identical bosons

    a   becomes zero

    b   changes sign

    c   acquires a factor $\gamma \neq 1$

    d   stays the same

7 Of the following, which are bosons?

    a α particles

    b   electrons

    c   helium-4 atoms

    d   helium-3 atoms

8 What will be the result of acting on a vacuum state by the following set of creation/annihilation Bose operators $c_5 (c_2)^2 (c_1^+)^5 (c_2^+)^3 c_5^+$ (up to a numerical factor)?

    a $|5,3,0,1,0...\rangle$

    b $|5,3,0,0,0...\rangle$

    c $|5,1,0,0,0...\rangle$

    d $|5,2,0,1,0...\rangle$

9 The Fock space is a mathematical construction, which

a Is a Hilbert space for a system of infinite number of particles

b replaces the Hilbert space for a system of identical quantum particles

c supplements the Hilbert space for a system of identical quantum particles

d is a special kind of Hilbert space suited for a system of identical quantum particles

10 The vacuum state is

a the lowest energy state of a quantum field

b the ground state of a quantum field

c the state from which all states can be obtained by the action of creation operators

d zero

Dig deeper

Further reading I

R. Feynman, R. Leighton and M. Sands, *Feynman Lectures on Physics*. Basic Books, 2010 (there are earlier editions). Vol. III, Chapter 4 (identical particles and the Pauli exclusion principle); Chapter 19 (the periodic table).

Further reading II. About the ideas and scientists.

J. Baggott, *The Quantum Story: A History in 40 Moments*, Oxford University Press, 2011. A more modern history. Well worth reading.

G. Gamow, *Thirty Years that Shook Physics: The Story of Quantum Theory*, Dover Publications, 1985. A history written by an eyewitness and one of the leading actors.

8

The virtual reality

Some of the most striking successes of quantum mechanics come from its application to the systems that contain a huge number of quantum particles. I mean condensed matter systems, like metals, dielectrics, semiconductors, superconductors, magnets and so on. These successes are also among those with the greatest impact on technology, economy and society.

One of the reasons for this success is the same as that of classical statistical mechanics. The behaviour of a large system is often determined by averages – average speeds, average densities, etc and these are much easier to measure and calculate than the properties of individual atoms or molecules. Finding averages in large quantum systems adds to this an extra layer: finding quantum expectation values. But we have already seen in Chapter 3 that switching from the state vector to the density matrix formalism can do this quite naturally.

Moreover, in this chapter we shall see that the description of a system of many quantum particles, which strongly interact with each other, can be often reduced to that of a system of almost non-interacting *quasi*particles with excellent results. It is the behaviour of a modest collective of quantum particles that are usually the hardest to describe and predict. There are exceptions, however. If quantum correlations involving many particles simultaneously are important, as is the case, for example, in quantum computing, these simplifications do not apply. For discussion of this situation please see Chapter 12.

Electrons in a jellium, charge screening, electron gas and Fermi–Dirac distribution

We begin with the simplest quantum theory of metals. The characteristic properties of metals – their high electrical and thermal conductivity, their metallic lustre, their strength, their ability to be melted or hammered or pressed into a desired shape without breaking apart – in short, everything that makes them useful is due to the fact that metals contain a large number of free electrons.

If look at Mendeleev's Periodic Table, we see that the most typical metals, the alkali metals (lithium, sodium, potassium, etc.) are in the same column as hydrogen – that is, they have one unpaired electron in their outermost s-shell. This electron is the closest to the top of the potential well and is therefore easily lost by the atom. This, in particular, explains the extraordinary chemical activity of the alkali metals.

Most other elements in the Table are metals that also lose some of their outermost electrons pretty easily. As a result, when atoms of these elements group together, these electrons *collectivize*. That is, they are no longer bound to any specific atom. Instead they can move anywhere in the piece of metal and, e.g. can carry electric current. This is why they are called *conduction electrons*. Other electrons, which occupy lower-energy states, remain with their nuclei and form the positively charged ions (Figure 8.1).

Though there is no individual bonding between ions and conduction electrons, they are bound to each other collectively. The two subsystems, positively charged ions and negatively charged conduction electrons, are strongly attracted by the electrostatic forces. This forms a specific *metallic bonding*. The electrons, while free to move about and repel each other, cannot leave the metal in any significant number, because then they would be attracted back by the strong electric field of the uncompensated ions. (A relatively tiny number of electrons can be taken from the metal – this is what happens in photoelectric

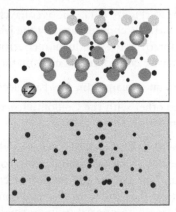

Figure 8.1 Metal (a) and its jellium model (b)

effect or in high-school demonstration experiments with electrostatic polarization. And, of course, electric current flow, however strong, does not change the number of electrons in the metal wire – as many leave as arrive per unit time.) The ions, in their turn, are held together by their attraction to the conduction electrons. And Coulomb repulsion prevents this substance from collapsing: the density of the material is, in the final account, determined by the equilibrium between the ion–ion (and electron–electron) repulsion and the ion–electron attraction.

This picture nicely explains why metals not only conduct electricity and heat (electrons can and do carry thermal energy as well), but are also malleable and strong. The ions and conduction electrons are held together by electrostatic forces, which are very powerful and (as long as we have a very large number of electrons and ions packed in a unit volume) are not very sensitive to the exact positions of these electrons and ions. Therefore a piece of metal can be bent without breaking; it can be melted and poured in a desired shape; two pieces of metal can be hammered or welded together; and so on.

We can go even further. The simplest model of a metal (the jellium model) disposes of the ions altogether. They are replaced by a uniformly distributed positive charge density – like a jellium – against the background of which move the conduction electrons (Figure 8.1). The average charge density of the electrons, $-|e|\,n_0$ is compensated by the jellium charge density $+|e|\,n_0$. Moreover, the conduction electrons, to a good approximation, can be considered as non-interacting.

The latter requires an explanation. Any two electrons in empty space repulse each other with the Coulomb force $F = \frac{e^2}{r^2}$. But a metal is *not* an empty space.

Suppose we have placed an external positive electric charge Q is inside a piece of metal and choose this point as the coordinate origin. The nearby conduction electrons will be attracted to it, so that the local electron density $n(r) = n_0 + \delta n(r)$, will increase compared to its normal value. The attraction force will decrease with the distance from the origin, but *faster* than the standard Coulomb $1/r^2$. The reason is simple. As conductance electrons gather at the positive charge, they form an additional negative charge in its vicinity. Within the distance R from the origin this charge is $\delta Q_R = -|e| \int_{V_R} d^3r \, \delta n(r)$ (Figure 8.2). An electron at the distance R from the origin will be attracted not to the *bare* charge $+Q$, but by the smaller *screened* charge $Q_R = Q - \delta Q_R$. The same

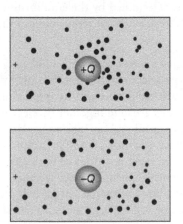

Figure 8.2 Charge screening in the jellium model

happens if the probe charge is negative. Now the conduction electrons will be repulsed and the positively charged jellium will screen the probe charge, with the same result.

Taking one of the electrons as the probe charge, we see that these very electrons also screen its interaction with other electrons. It acts over a short distance only, and it is therefore plausible that the conduction electrons in a metal behave as a *gas* – the *electron gas*. (Recall that classical statistical physics treats a gas as a system of particles, which only interact when colliding with each other.)

This is not quite a usual gas however. Electrons are fermions, and their *distribution function* $n_F(E)$ (the number of electrons with energy E) is strongly affected by the fact that they must obey the Pauli principle. Therefore instead of the familiar Boltzmann distribution we have the *Fermi–Dirac distribution*:

$$n_F(E) = \frac{1}{\exp\frac{E - \epsilon_F}{K_B T} + 1}$$

Spotlight: Enrico Fermi

Enrico Fermi (1901–1954) was an Italian physicist, and one of the few great physicists who excelled equally in theory and experiment.

Fermi applied the Pauli exclusion principle to the statistical mechanics of an electron gas and came up with the Fermi–Dirac statistics (so-called because Dirac obtained the same results independently). Later Fermi developed the first successful theory of weak interactions; it included the light neutral particle, proposed earlier by Pauli, which Fermi called *neutrino* (a little neutron).

His Nobel Prize (1938) was awarded for his experimental research in radioactivity and he played an important role in the development of nuclear weapons and nuclear power.

Fermi's textbook on thermodynamics is a classic and still one of the best ever written.

Key idea: Thomas–Fermi screening

In the case of high density of conduction electrons (which in good metals is of order $n_0 = 10^{22}$ cm^{-3}), one can see if the electrostatic potential $\Phi(\mathbf{r})$ of the probe charge Q satisfies the *Poisson's equation*:

$$\nabla^2 \Phi(\mathbf{r}) = \frac{1}{\lambda_{TF}^2} \Phi(\mathbf{r}).$$

Its solution is given by

$$\Phi(\mathbf{r}) = \frac{Q}{r} e^{-\frac{r}{\lambda_{TF}}}.$$

This function drops exponentially fast, at the characteristic radius given by the *Thomas–Fermi length*

$$\lambda_{TF} = \frac{1}{|e|} \sqrt{\frac{\epsilon_F}{6\pi n_0}}.$$

The *Fermi energy* ϵ_F is the key characteristic of the electron gas and is usually of the order of a few electron volts.

The *Fermi energy* ϵ_F plays a very important role. To see this, let us look at the Fermi distribution in the limit of absolute zero (temperature $T = 0$). If $(E - \epsilon_F)$ is positive, the exponent in the denominator will go to infinity as $k_B T \to 0$. Therefore the function itself will be zero. On the other hand, if $(E - \epsilon_F)$ is negative, the exponent in the denominator will go to zero as $k_B T \to 0$, the distribution function will be equal to unity. Therefore at absolute zero the Fermi distribution is a sharp step (Figure 8.3),

$$n_F(E < \varepsilon_F; T = 0) = 1; \; n_F(E > \varepsilon_F; T = 0) = 0,$$

or, using the Heaviside step function, $n_F(E; T = 0) = \theta(\epsilon_F - E)$.

At a finite temperature, the step is smoothed out and acquires a finite width of order $k_B T$ (Figure 8.3). In a typical metal the Fermi energy is the order of a few electron volts; 1 eV corresponds to a temperature of approximately 12,000 K (twice the temperature

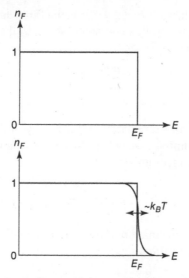

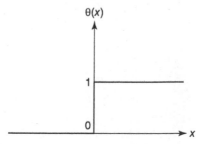

Figure 8.3 Fermi distribution function at zero and finite temperature

of the surface of the Sun). A metal will evaporate long before the thermal broadening of the Fermi step will be comparable to the Fermi energy. Therefore all the thermal processes in metals only involve a very small fraction of all conduction electrons that are present. This, by the way, provides an explanation to some puzzling features of the thermodynamic properties of metals, for which classical statistical physics could not account.

Key idea: Heaviside step function

A function $\theta(x)$, which is zero for all negative and one for all positive values of x is called the *Heaviside step function* (because he was the first to introduce it). Its derivative is the Dirac delta function (an infinitely sharp pulse with a unit area), which, by the way, also was first introduced by Heaviside:

$$\frac{d\theta(x)}{dx} = \delta(x).$$

Fermi energy is directly related to the density of the electron gas. Consider the simplest case, when the energy of an electron is related to its momentum via

$$E(p) = \frac{p^2}{2m}.$$

At absolute zero the maximal momentum an electron can have is the *Fermi momentum*,

$$p_F = \sqrt{2m\epsilon_F}.$$

Let us now calculate the electron density in a metal at absolute zero and in the absence of external fields. (The electrostatic field of the positively charged 'jellium', which holds the electrons in the metal, does not count.) Denote by Ω the volume of metal. In the one-electron phase space (Figure 8.4) an electron can have any position inside this volume, and any momentum p (smaller than p_F). The latter is because of the step-like Fermi distribution function $n_F(E)$. Since no two electrons can be in the same

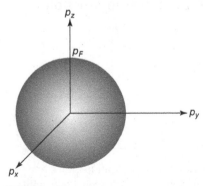

Figure 8.4 Fermi sphere

state, we can represent all electron states on the same picture: electrons *fill the Fermi sphere*.

Spotlight: Electron density calculation

As an exercise, calculate the electron density in aluminium (ϵ_F =11.7 eV), gold (ϵ_F = 5.53 eV), and lead (ϵ_F = 9.47 eV) (1 eV = 1.60 × 10⁻¹⁹ J). Use the mass of a free electron. Compare to the table values (respectively, 18.1 × 10²⁸, 5.90 × 10²⁸, 13.2 × 10²⁸ per cubic metre). What conclusions can you make?

The number of electrons is given by the occupied volume of the single-electron phase space divided by the Planck constant h^3 (to take into account quantization). This volume is the 'coordinate volume' Ω times the 'momentum volume' $\frac{4}{3}\pi p_F^3$. Therefore the electron density is

$$\frac{N}{\Omega} = 2 \times \frac{1}{h^3} \times \frac{4}{3}\pi p_F^3 = 2 \times \frac{4\pi}{3h^3}(2m\epsilon_F)^{\frac{3}{2}}.$$

The factor of two is due to the electron spin: two electrons with opposite spins can occupy each state.

In a more general case (for example, at a finite temperature) the electron density is given by the expression

$$\frac{N}{\Omega} = \frac{2}{h^3}\int d^3\boldsymbol{p}\, n_F(E(\boldsymbol{p})).$$

Key idea: Density of states

The *density of states* is the number of electron states per unit energy (and unit volume). It is given by the derivative of the number of electron states with energies less or equal to E, $\nu(E)$:

$$D(E) = \frac{d}{dE}\frac{\nu(E)}{\Omega}.$$

In the simplest case $D(E) = \frac{d}{dE}\left(2 \times \frac{4\pi}{3h^3}(2mE)^{\frac{3}{2}}\right) = \frac{4\pi}{h^3}(2m)^{\frac{3}{2}}E^{\frac{1}{2}}.$

Electrons in a crystal: quasiparticles*

We have just 'considered the simplest case', when the electron energy was related to its momentum as $E(p) = \frac{p^2}{2m}$. But what other cases may there be?

As a matter of fact, a conduction electron practically never has such a *dispersion law* (the relation between the kinetic energy and the momentum). The mass in the above expression can be as large as about 1000 free electron masses (in some compounds such as $CeCu_6$, $CeAl_3$, UBe_{13} or UPt_3) or as small as 0.044 m_e (in germanium). Therefore physicists usually write m^*, the *effective mass* of an electron, in the above formulas.

Worse still, the 'mass' can be anisotropic, so that the energy is related to the momentum through an 'inverse mass tensor':

$$E(p) = \frac{1}{2} \sum_{j,k=x,y,z} m_{jk}^{-1} \, p_j p_k.$$

Here is the simplest example of such dependence:

$$E(p) = \frac{p_x^2}{2m_{xx}} + \frac{p_y^2}{2m_{yy}} + \frac{p_z^2}{2m_{zz}}.$$

The electron will behave as if it had different masses when pushed along different axes.

To make things even more interesting, there can be several different sorts of 'electrons' at once, some of them can even have positive electric charge.

All this happens even if we completely neglect the interaction of electrons with each other and only consider their interaction with the ions. And much of what happens is due to the fact that ions form an ordered crystal lattice. (This is where the jellium model fails.)

Electrons are quantum objects and are described by wave functions. We have already seen in Chapter 6 that an electron scattered by a double slit or a diffraction grating (the role of which was played by the rows of atoms on the surface of a material, e.g. nickel) produces an interference pattern. The same happens when a conduction electron moves through the lattice of positive ions, which are arranged in a regular way: it is scattered. As the result, the electron wave function will produce an interference pattern, and the expectation value of momentum and energy will be affected compared to what one would expect for an electron wave in the free space. This is not surprising, after all, since the electrons are acted upon by the ions. The ions themselves are much heavier than electrons and the reciprocal action of electrons on the ions can be usually neglected.

Since conduction electrons in a metal (or in a semiconductor) usually have properties very different from those of a free electron in space, they are called *quasiparticles*.

Conduction electrons are not the only quasiparticles in solids (Figure 8.5). There are such quasiparticles as phonons, polaritons, magnons, spinons, polarons, etc. Unlike conduction electrons, some of them cannot be brought to the outside. They only exist within, like the units in a MMPORG (a massively multiplayer online role-playing game). For example, a phonon is a quantum of sound. But in the solid it has all the properties of a quantum particle, with its own energy, momentum, dispersion relation and quantum statistics (phonons, by the way, are bosons, not fermions).

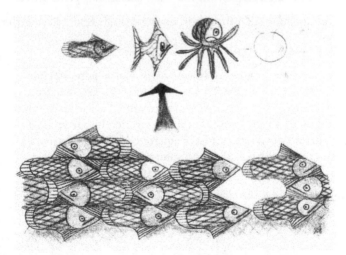

Figure 8.5 Quasiparticles (Zagoskin 2014: 55, Fig. 2.1)

Electrons in a crystal: Bloch functions; quasimomentum; Brillouin zones; energy bands; and Fermi surface*

But let us go back to conductance electrons in a periodic lattice. The remarkable fact is that the very presence of a periodic scattering potential, no matter how weak, already changes the properties of a free electron. This is expressed by *Bloch's theorem* (due to the same Felix Bloch of Chapter 4), according to which any solution of the stationary Schrödinger equation

$$-\frac{\hbar^2}{2m}\nabla^2\Psi(r) + U(r)\Psi(r) = E\,\Psi(r)$$

with a periodic potential $U(r)$ has the form of a *Bloch wave*:

$$\Psi_p(r) = u_p(r)e^{ip.r/\hbar}.$$

Here the vector p is called the *quasimomentum* and depends on the energy E, and the amplitude $u_p(r)$ has the same periodicity as the crystal potential $U(r)$. The wave function is thus a periodically modulated (but not itself periodical) plane wave.

Periodicity means that there exist one or several vectors R, such that

$$U(r + qR) = U(r)$$

for any integer number q. In other words, the potential has *translational symmetry*. There is a beautiful relation between translational and rotational symmetry, the discussion of which would take us too far away from our topic. (In particular, it explains why crystals can have square or hexagonal symmetry, but not pentagonal – and why something that looks like a crystal with a pentagonal symmetry is not truly periodic.)

We will consider the simplest possible example: a one-dimensional periodic lattice. The solution of the Schrödinger equation in the periodic potential $U(x + X) = U(x)$ will have the form of a Bloch function, $\Psi_p(x) = u_p(x)e^{\frac{ipx}{\hbar}}$. Note that this wave function is not itself periodic:

$$\Psi_p(x + X) = u_p(x + X)e^{\frac{ip(x+X)}{\hbar}} = u_p(x)e^{\frac{ipx}{\hbar}}e^{\frac{ipX}{\hbar}} = \Psi_p(x)e^{\frac{ipX}{\hbar}}.$$

This does not affect the periodicity of observable quantities, since the phase factor $e^{\frac{ipx}{\hbar}}$ is cancelled when we calculate, e.g. the probability of finding an electron at a given point:

$$\left|\Psi_p(x + X)\right|^2 = \left|\Psi_p(x)\right|^2 \left|e^{\frac{ipX}{\hbar}}\right|^2 = \left|\Psi_p(x)\right|^2.$$

But the phase factors uniquely label different Bloch functions. Then we see that any two values of momentum, which differ by $G = 2\pi\hbar/X$, should correspond to *the same* Bloch function, because they acquire the same phase factors after a translation:

$$\Psi_p(x + X) = \Psi_p(x)e^{\frac{ipX}{\hbar}},$$

while $\Psi_{p+G}(x + X) = \Psi_{p+G}(x)e^{\frac{ipX}{\hbar}}e^{\frac{iGX}{\hbar}} = \Psi_{p+G}(x)e^{\frac{ipX}{\hbar}}e^{2\pi i}$

$$= \Psi_{p+G}(x)e^{\frac{ipX}{\hbar}}.$$

Of course, the same holds for any integer multiple of G. Therefore the variable p, which would be the uniquely defined momentum of an electron wave in free space, is now defined *modulo G* (that is, p and $p+sG$ are equivalent for any integer s) and, quite logically, is called *quasi*momentum.

Let us consider the consequences in the 'empty lattice model' – that is, take a free electron and look *only* at the effects of replacing the momentum with the quasimomentum.

The electron energy is $E(p) = \frac{p^2}{2m}$. Since p is now only defined modulo G, we can only consider the interval of length G, for example, take

$$-\frac{G}{2} \leq p < \frac{G}{2}.$$

(This interval is called the *first Brillouin zone*.) In order to fit the entire energy spectrum into this zone, we will have to 'fold' it (see Figure 8.6). Now there are several (in this case, infinitely many) energy values corresponding to the same quasimomentum: $E_1(p)$, $E_2(p)$,... These are the *energy bands*. They are due to the scattering of the electron wave by the periodic lattice formed by the ions.

In case of a two- or three-dimensional lattice the result will be similar. Instead of a single G there will be respectively two or three vectors[1], so that, for example, in three dimensions the quasimomenta p and $p + s_1 G_1 + s_2 G_2 + s_3 G_3$ are equivalent for any integer numbers s_1, s_2, s_3. The first Brillouin zone will have a more complex shape, depending on the symmetry of the lattice.

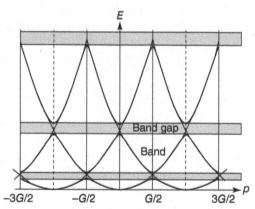

Figure 8.6 Electron in an empty 1D lattice: bands and band gaps

The empty lattice model is quite useful in helping visualize at least a rough outline of the energy bands imposed by the symmetry of the lattice. The actual picture will be, of course, different and can be determined only by numerical simulations. In the first place, at the zone boundary the

---

1 They are called the inverse lattice vectors, but you do not have to remember this.

derivative $\frac{dE}{dp} = 0$. Therefore a gap (the *band gap*) opens in the spectrum: some values of energy for a conduction electron are not allowed.

In a real 3D lattice the resulting band structure can be very intricate, but the principle is the same.

Key idea: Bloch oscillations

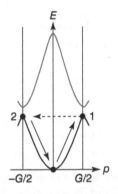

The consequences of periodicity are pretty serious. Suppose an electron is accelerated by an external field. As its quasimomentum increases, and it reaches the zone boundary (1), it will reappear on the other side, with quasimomentum $p - G$ (2), and the process will repeat itself. (Usually such *Bloch oscillations* are prevented from happening by other processes, e.g. collisions of an electron with phonons.)

The band theory of solids explains, among other things, why some materials are metals, some are semiconductors, and some are dielectrics. In the first place, this is determined by their band structure and the electron density (that is, the Fermi energy). Suppose the Fermi energy is in a band gap (Figure 8.7). Then all the lower bands are occupied and all the upper bands are empty. There is no current in the system: there

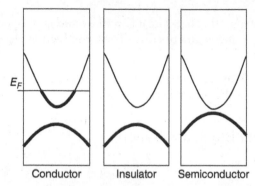

Figure 8.7 Conductors, semiconductors and dielectrics (alias insulators)

is the same number of electrons with negative as with positive quasimomenta.

If we apply a small electric field the current will not flow, since the electrons will be unable to move: all available states are already occupied. Our system is a dielectric.

The situation is different if the Fermi energy is above the bottom or below the top of a band. In either case there are states that are available, and a small electric field will produce the imbalance, that is, the net current. This is the case of a metal.

You see that the behaviour of the system depends on the level to which the electrons are 'poured' into it. (This is why the Fermi energy is also often called the *Fermi level*.)

At a finite temperature the Fermi level becomes 'fuzzy': the sharp Fermi step is smeared in the interval of order $k_B T$. Therefore, if in an insulator the uppermost band (*valence band*) is close enough the next empty band (*conduction band*), there will be some thermally excited electrons in the conduction band, and empty states (*holes*) in the valence band. If we apply an electric field, both will move and carry current. This is what happens in semiconductors.[2]

---

2 Usually semiconductors are *doped* to increase their electrical conductivity. That is, certain impurity atoms are put in the material, which either add extra electrons to the conductance band or take some electrons from the valence band (and thereby create holes there).

Note that in the presence of electric field electrons and holes will move in the opposite directions, but carry the electric current in the same direction.

In a metal the surface in the quasimomentum space, determined by the condition $E(p) = \varepsilon_F$, is called the *Fermi surface*. In the jellium model it was simply a sphere of radius p_F. In real metals it can be much more interesting (Figure 8.8). The shape of the Fermi surface and the band structure to a large degree determine what kind of quasiparticles exist in the system.

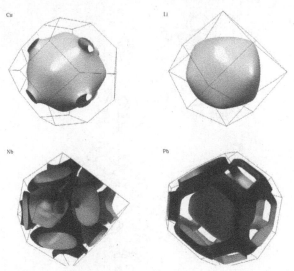

Figure 8.8 Fermi surfaces (copper, lithium, niobium, lead) (University of Florida 2007)

Quasiparticles and Green's functions

The properties of conductance electrons are affected not only by their interaction with the ions of the crystal lattice, but also by their interactions with each other and with other quasiparticles in the system. Consider, for example, the screening of electrostatic potential in the jellium model, which we discussed at the start of this chapter. If a charged particle moves through the system, its potential is screened by the electrons, which will shadow it (forming a region with either positive or negative extra electron density, depending on whether the particle is positively or negatively charged). Besides changing the shape of the effective potential of the particle, there will be two other effects. First, in its motion the particle drags along a 'cloak' of electrons (the technical term is *'dressed'*). Therefore its mass will be effectively larger than in the absence of such a 'cloak' (i.e. than the mass of the *bare* particle). Second, if the particle moves fast enough, the 'cloak' will not have time to form, and both the effective mass and the electric field of the particle will become closer to their bare values (Figure 8.9).

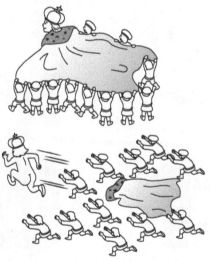

Figure 8.9 Dressed and bare particles

This is a very general phenomenon called *renormalization*. In some form it takes place in every many-particle system. The difficulty of dealing with it lies in the necessity to take into account all the interactions in their dynamics. In other words, one would have to solve the Schrödinger equation for all the particles in the system at once. This is a task even more impossible than solving Hamilton's equations for all the particles in an interacting classical gas.

Fortunately, there are ways around this obstacle. Many properties of a physical system can be obtained from its *response* to an external perturbation – like knocking on a wall to discover a hidden cavity. One way of perturbing a system of quantum particles is to add a particle at some point, wait and see what happens, and then take it away at some other point. Alternatively, we can instead remove a particle, wait, and then replace it (Figure 8.10).

In the first case, the quantum state vector of the system will be

$$|\Phi\rangle = \hat{\psi}(r', t')\hat{\psi}^{+}(r, t)|\Psi\rangle$$

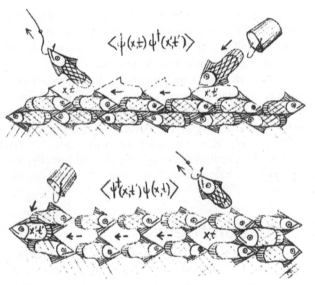

Figure 8.10 Green's functions (Zagoskin 2014: 58, Fig. 2.3)

(we have first used the creation field operator $\hat{\psi}^{+}(r,t)$ to create a particle at the point r at the moment t, and then used the annihilation field operator $\hat{\psi}(r', t')$ to annihilate a particle at r', t'). The effect of such a manipulation can be evaluated if we calculate the scalar product of the post-factum state $|\Phi\rangle$ with the initial state $|\Psi\rangle$. (The closer it is to unity, the smaller is the effect.) This scalar product, $\langle\Phi|\Psi\rangle$, is (up to a factor) a *Green's function*:

$$\langle \Phi | \Psi \rangle = \langle \Psi | \hat{\psi}(r', t')\hat{\psi}^+(r,t) | \Psi \rangle = \langle \hat{\psi}(r', t')\hat{\psi}^+(r,t) \rangle_\Psi$$
$$= -iG^{+-}(r', t'; r, t).$$

As you can see, it can be also regarded as the expectation value of the product of two field operators, taken in the state $|\Psi\rangle$ of our quantum system.

Another Green's function corresponds to the second way of probing our system:

$$\langle \tilde{\Phi} | \Psi \rangle = \langle \Psi | \hat{\psi}^+(r', t')\hat{\psi}(r,t) | \Psi \rangle = \langle \hat{\psi}^+(r', t')\hat{\psi}(r,t) \rangle_\Psi$$
$$= \pm iG^{-+}(r', t'; r, t).$$

Do not pay much attention to the tildes, pluses and minuses – there are several kinds of Green's functions, defined in a similar way, labelled with different sub- or superscripts and with different signs (which also depend on whether the field operators are bosons or fermions). The imaginary unit is introduced for mathematical convenience. What really matters though is: (i) that the important physical properties of a quantum system can be directly expressed through these Green's functions[3]; and (ii) in many important cases the Green's functions can be calculated in a systematic and literally very graphic way. This makes Green's functions the mainstay of quantum theory of many-body systems.

Perturbation theory

Very few problems in physics have an exact solution. This is not to say that all such problems have already been discovered and solved, but it is far more probable that any new problem does not have an exact solution. Numerical methods are helpful and powerful, but despite the ever growing computing power there are many problems they cannot solve now and will never solve because of the sheer scale of required computations (see the

---

3 For example, you can check that the average particle density in the system is given by $\pm iG(r,t;r,t)$. .

discussion in Chapter 12). Moreover, numerical computation is akin to experimental physics: it does not necessarily give insight into the key features of the effects that are being modelled (wherefore the term *computer experiment*). Numerical computations are also vulnerable to flaws in the underlying computational models, and without some guidelines obtained in other ways it is hard to distinguish model-dependent artefacts from real effects.[4] It is always good to have some guidelines against which to measure the numeric, and which can provide insight into the mechanism of an effect and its likely dependence on the model parameters. In short, physicists still prefer to have analytical solutions, even in the situation when exact solutions are not known or do not exist.

The way of dealing with this situation was first found in Newton's time, and it has served as the mainstay of theoretical physics ever since. I am talking about the *perturbation theory*. The idea of the approach is very simple. If our problem does not have an exact solution, then we will look for a similar problem that has. We then will modify that solution to fit our problem. There are systematic ways of doing it, and they are very useful. Laplace very successfully applied it to the calculation of planetary motions in his famous work on celestial mechanics, published long before the appearance of computers.

In quantum mechanics, the perturbation theory works if the Hamiltonian of the system can be presented as a sum of the 'main term' and the perturbation:

$$\hat{H} = \hat{H}_0 + \epsilon\hat{H}_1.$$

---

4 One of the key problems of climate science is that equations of hydrodynamics, on which it is essentially based, are some of the most difficult in all of physics and for realistic systems can only be tackled numerically. Add to this the paucity of data on which to base and with which to compare the calculations, the huge scale of the system, a number of unknown processes or parameters, and you understand why making any confident predictions in this field is such a risky business.

Case study: Perturbation theory in classical mechanics: a cubic oscillator

The equation of motion for a harmonic (or linear) oscillator (like a mass on a spring, when the elastic force satisfies the Hooke's law)

$$m\frac{d^2x}{dt^2} = -kx,$$

has simple exact solutions (for example, $x_0(t) = A \sin \omega t$, where $\omega^2 = k/m$).

In reality springs are not ideal, and there are deviations from Hooke's law, for example:

$$F(x) = -kx - k\epsilon x^3.$$

As long as this *cubic nonlinearity* is small, $\epsilon \ll 1$, we can use the perturbation theory. To do so, let us present the solution as a series in powers of ϵ:

$$x(t) = x_0(t) + \epsilon x_1(t) + \epsilon^2 x_2(t) + \dots$$

Since $\epsilon \ll 1$ the corrections will be small, and only the main terms need to be kept. Substituting this expansion in our equation, you will find:

$$m\epsilon\frac{d^2x_1}{dt^2} = -k\epsilon x_1(t) - k\epsilon(x_0(t))^3 + \dots$$

The remaining terms will be multiplied by ϵ^2, ϵ^3, etc. and are therefore much smaller. Dividing by ϵ we find for the first correction the equation

$$m\frac{d^2x_1}{dt^2} = -kx_1(t) - k(A \sin \omega t)^3.$$

This equation can be solved explicitly and yield an approximate solution for the cubic oscillator. If better accuracy is required, we look at the terms with higher powers of ϵ. There will be some subtle problems with this expansion, but they also can be taken care of in a similarly systematic way.

Here I explicitly include a small parameter $\varepsilon \ll 1$. The state vector of the system will be sought in the form of a power series:

$$|\Psi\rangle = |\Psi_0\rangle + \epsilon|\Psi_1\rangle + \epsilon^2|\Psi_2\rangle + \ldots + \epsilon^n|\Psi_n\rangle + \ldots$$

As $\epsilon \to 0$, we will need a smaller and smaller number of terms in this series to obtain an accurate answer.

Now we substitute this state vector in the Schrödinger equation, $i\hbar\frac{d}{dt}|\Psi\rangle = \hat{H}|\Psi\rangle$, and collect the terms with the same power of ϵ:

ϵ^0	×	$i\hbar\dfrac{d}{dt}|\Psi_0\rangle = \hat{H}_0|\Psi_0\rangle$
ϵ^1	×	$i\hbar\dfrac{d}{dt}|\Psi_1\rangle = \hat{H}_0|\Psi_1\rangle + \hat{H}_1|\Psi_0\rangle$
ϵ^2	×	$i\hbar\dfrac{d}{dt}|\Psi_2\rangle = \hat{H}_0|\Psi_2\rangle + \hat{H}_1|\Psi_1\rangle$
...	...	...

We can solve them one after another. Presumably we already know the solution $|\Psi_0\rangle$ to the first equation (otherwise we would have chosen a different $\hat{H}_0$). Then we can substitute it in the second equation and find from there $|\Psi_1\rangle$. Then we find $|\Psi_2\rangle$, and so on.

The procedure is not as easy as one could think, especially if we deal with a many-body system. Fortunately, if we are interested in Green's functions and not in the state vector itself, the perturbation theory is greatly simplified and can be represented in the form of *Feynman diagrams*.

Feynman diagrams and virtual particles

Feynman rules allow us to draw diagrams and then translate them into explicit mathematical expressions, which correspond to different terms in the perturbation expansion of a Green's function. What is even better is that these diagrams have a certain appeal to our intuition (although intuition in quantum mechanics should be used with great caution).

Let us start with fermions. Green's function[5] can be looked at as a way to describe the propagation of a fermion from the point r at the moment t to the point r', where it arrives at the moment t'. We will therefore draw for it a thick solid line with an arrow, and mark the ends appropriately (Figure 8.11). We assume that time runs from the right to the left.

	$iG(r', t'; r, t)$	Green's function (fermions)
	$iG_0(r' - r, t' - t)$	Unperturbed Green's function (fermions)
	$iD(r', t'; r, t)$	Green's function (bosons)
	$iD_0(r' - r, t' - t)$	Unperturbed Green's function (bosons)
	$-iV(r' - r)$	Interaction potential
	$-ig$	Fermion-boson coupling

Figure 8.11 Feynman rules

If there are no external forces, no interactions, and the system is uniform and at equilibrium, the Green's function can be calculated explicitly (for example, in the jellium model, neglecting the electron-electron repulsion). Such function is denoted by $G_0(r' - r, t' - t)$ – indeed, in a uniform and stationary system there are no special points in space or special moments in time. Everything can only depend on the displacements and time differences. We will draw a thin solid line for this function.

The difference between the unperturbed Hamiltonian $\hat{H}_0$ and the actual Hamiltonian $\hat{H}$ is in the first place due to the

---

5 This will be a certain combination of the functions G^{+-} and G^{-+}, but we shall not discuss such details here.

interactions between the particles. We will again consider this against the background of the jellium model of a metal.

First, there is an electrostatic interaction between electrons. The corresponding perturbation term in the Hamiltonian is

$$\hat{H}_c = \int d^3r' \int d^3r \, \hat{\psi}^+(r',t)\hat{\psi}(r',t)V(r'-r)\hat{\psi}^+(r,t)\hat{\psi}(r,t).$$

It has a transparent physical meaning: an electron at the point r' interacts with an electron at the point r with the potential $V(r'-r) = e^2/|r'-r|$. The probability densities of either electron being there are $\langle\hat{\psi}^+(r',t)\hat{\psi}(r',t)\rangle$ and $\langle\hat{\psi}^+(r,t)\hat{\psi}(r,t)\rangle$ respectively. In the diagram the potential will be denoted by a broken line (Figure 8.11).

Second, there is interaction between electrons and phonons (sound quanta). In other words, the ions in the crystal lattice can oscillate near their equilibrium positions, and this affects the electrons. In the jellium model, the 'ion jelly' wobbles a little. These quantized wobbles – phonons – are usually described by *Hermitian* bosonic field operators (as we have seen in Chapter 7), and their interaction with electrons can be taken into account by adding to the Hamiltonian the term

$$\hat{H}_{ph} = g\int d^3r \, \hat{\psi}^+(r,t)\hat{\varphi}(r)\hat{\psi}(r,t).$$

The Green's functions of bosons (such as phonons or photons) usually do not depend on the direction in which the boson propagates. This is related to the fact that the Bose field operators can be made Hermitian, as we have seen in Chapter 7 – and in the expression like $\langle\hat{\varphi}(r',t')\hat{\varphi}(r,t)\rangle_\Psi$ how can you tell, which is the 'creation' and which the 'annihilation' end?

In order to clearly distinguish the bosonic from the fermionic Green's functions, we will draw for them wavy lines without arrows and denote them by $D(r',t';r,t) = -i\langle\hat{\varphi}(r',t')\hat{\varphi}(r,t)\rangle_\Psi$ and $D_0(r'-r,t'-t)$ (Figure 8.11).

Now we can finally start drawing pictures (Figure 8.12).

The rules are simple:

▶ Green's function is the sum of contributions in all orders of the perturbation theory.

▶ Each contribution is a diagram and only different diagrams should be included.

▶ Each diagram must have only two loose ends: the ones that correspond to the 'in' and 'out' ends of Green's function are what we are trying to find.

▶ Fermion lines cannot end inside a diagram. (This reflects an important conservation law: fermions can only appear and disappear in pairs.)

▶ Boson lines can. (For example, an electron can emit a phonon or a photon.)

▶ No diagram should fall apart (that is, contain disconnected parts). (This rule follows from the way the perturbation series is derived.)

The zeroth order contribution is simply G_0.

The first order contains diagrams with one interaction vertex (that is, V or g). You can easily see that such diagrams will have extra loose ends, and therefore there is no first-order contribution.

The second order will contain four diagrams (Figure 8.12). As promised, they speak to our intuition. For example, (a) an electron on its way from A to B interacted with a 'bubble' via electrostatic potential at some point C. The 'bubble' is actually the unperturbed density of electrons – and we started taking into account the electron's interaction with other electrons, that is, the screening of the Coulomb potential. When calculating the mathematical expressions, corresponding to each diagram (which we most certainly will *not* do here), we would integrate over all the internal coordinates and times, over all space and all times. Therefore the outcome will not depend on where and when the 'act of interaction' takes place.

The diagram (b) is more interesting. Since the interaction is instantaneous, we draw the potential line vertically. We are forced then to draw the internal electron line in a backward

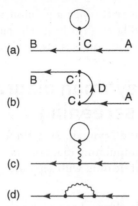

Figure 8.12 Second order Feynman diagrams for the electron's Green's function

curve, formally making it propagate back in time. A better way of looking at this is to think that at point D a particle-antiparticle pair was created. Then the antiparticle annihilated with the original electron at C, and the particle interacted at the point C' and moved on. Since all quantum particles of the same kind are undistinguishable, this makes no difference.

The 'antiparticle' in our model is a hole – an empty state below the Fermi surface. We can consider the lowest energy state of our system – the filled Fermi sphere – as the vacuum state. Note that the temporary creation of a pair out of a vacuum does not violate any conservation laws, because this is a pair of *virtual particles*.

Virtual particles are, in a sense, artefacts of the perturbation theory: they correspond to the internal lines of Feynman diagrams. But they do provide a useful picture if one remembers the key difference between virtual and real particles.

For a real particle (or, as in a metal, a real *quasi*particle, like the conduction electron) there is a fixed relation between its energy and momentum (the dispersion law). For example, $E(p) = p^2/2m$. For a virtual (quasi)particle there is no such constraint. It can have any energy and any momentum, because it cannot escape from the diagram, so to speak.

For example, the bosons – phonons – in the diagrams (c) and (d) are virtual. In the diagram (c) an electron interacts with the average electron density through a phonon, and in (d) it emits a virtual phonon and later absorbs it. The electron between the emission and absorption points is also a virtual particle.

Summing Feynman diagrams and charge screening

The main power of the Feynman diagram technique lies in the possibility of manipulating diagrams almost as if they were numbers and of dealing with separate parts of a diagram separately.

Consider, for example, the problem of the electric charge screening (see 'Electrons in a jellium' earlier on in this chapter). The 'bare' Coulomb interaction is depicted by a thin dotted line. Now, for *any* diagram that contains this line, the perturbation theory expansion will also include diagrams that contain a line with an 'electron-hole loop' (the electron and hole are, of course, virtual), two loops, three loops and so on *ad infinitum* (Figure 8.13).

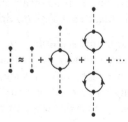

Figure 8.13 Dressing the bare potential

We can therefore sum these diagrams separately and obtain the 'dressed' potential (the thick dotted line), and then use it instead of the 'bare' one. The summation itself can be conducted if we notice that the series of 'loopy' diagrams is a geometric progression, which is drawn in Figure 8.13 and can be *symbolically* written as[6]

$$V_{eff} \approx V + V * \Pi_0 * V + V * \Pi_0 * V * \Pi_0 * V + \ldots = V + V * \Pi_0 * V_{eff}$$

---

6 We use here an approximate equality, because there are other diagrams contributing to V_{eff}.

(the star here stands for a certain set of mathematical operations, like integrations over positions and momenta of the virtual particles). The solution is

$$V_{eff} \approx \frac{1}{1 - V * \Pi_0} * V.$$

Here the fraction is also to be understood symbolically, but the necessary mathematical operations, though more complicated than just calculating an inverse of a number, are well defined. If we carry them through in the case of a jellium model, we obtain the Thomas–Fermi screened potential at the beginning of this chapter.

Dressing the quasiparticles

Now we can finally tell how the interactions change the properties of a quantum particle. Take the conduction electron. The expansion of its Green's function can be drawn and written as follows (Figure 8.14):

$$G = G_0 + G_0 * \Sigma * G_0 + \cdots = G_0 + G_0 * \Sigma * G.$$

Here the *self-energy* Σ is the sum of all diagrams, which have one 'input' and one 'output' and cannot be cut in two by separating just one electron line. They are therefore put on the

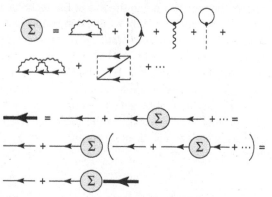

Figure 8.14 Green's function of a particle and self-energy

electron line as beads on a string. The symbolic solution will be like that for the effective potential:

$$G = \frac{1}{1 - G_0 * \Sigma} * G_0.$$

If written explicitly, we would see that the mass of a dressed conductance electron would change, as well as its other properties. But what is most important is that we can still consider our system as consisting of *independent* quasiparticles: once interactions are taken into account in dressing electrons, we can forget about them, at least, as long as the electron energies are close enough to the Fermi energy. Now recall that due to the sharpness of the Fermi distribution only electrons with such energies are important anyway.

We have achieved a huge simplification: instead of explicitly dealing with astronomical numbers of interacting real electrons and ions we can now be concerned with almost independent quasielectrons and phonons (and other quasiparticles, which may exist in our system). On our way we have also made some observations – such as the difference between bare and dressed particles and interactions – which reach far beyond the field of quantum theory of solids.

Spotlight: 'The Green Monster'

The first systematic monograph on the use of the techniques we have discussed here (and many other things) in condensed matter systems, *Quantum Field Theoretical Methods in Statistical Physics* by A.A. Abrikosov, L.P. Gor'kov, and I. Ye Dzyaloshinskii, was published in Russian in 1962 and in English in 1965. Its Russian edition was popularly known as 'the Green Monster' – in all probability, because of the proliferation of Green's functions in the text and formulas, the novelty and complexity of the subject, the not exactly user-friendly style, and the colour of its cover.

Fact-check

1 A metal is held together by

   a   strong forces

   b   electromagnetic forces

   c   electrostatic attraction between conduction electrons and lattice ions

   d   gravity

2 The screening of an electric charge in metal is due to

   a   long-range character of electrostatic interaction

   b   positive charge of the lattice ions

   c   periodicity of lattice

   d   presence of free conduction electrons

3 The Fermi energy of an electron gas is

   a   the energy of a fermion

   b   the maximal energy which can have an electron in the electron gas

   c   the maximal energy of an electron in equilibrium at absolute zero

   d   the characteristic broadening of the Fermi distribution function

4 Electrons in a crystal lattice are called quasiparticles because

   a   their electric charge can be fractional

   b   they do not interact with each other

   c   they interact with each other

   d   their properties are different from those of electrons in vacuum

5 According to the band theory, conductors conduct electricity, because

   a   they contain electrons

   b   their lowest energy bands are filled

   c   their uppermost energy band is filled

   d   their uppermost energy band is partially filled

6 A dressed quasiparticle differs from a bare quasiparticle by
 a a coat of other particles interacting with it
 b a strong interaction with the lattice
 c the electric charge
 d the quasimomentum

7 A Green's function is
 a an expectation value of a product of two field operators
 b a scalar product of two quantum state vectors
 c a measure of a many-particle quantum system's response to an external perturbation
 d the quantum state of a system of quasiparticles

8 A Feynman diagram is
 a a graphic representation of a quantum system
 b a graphic representation of a state of a quantum system
 c a graphic representation of one term in the perturbation expansion of a Green's function
 d a graphic representation of a series of terms in the perturbation expansion of a Green's function

9 A virtual particle
 a cannot have momentum, energy or electric charge
 b is represented by an internal line in a Feynman diagram
 c is a quantum particle, which exists for a limited time
 d does not satisfy the relation between the momentum and energy

10 An antiparticle or a hole is represented by
 a a solid line
 b a wavy line
 c a line running backwards in time
 d a bubble

Dig deeper

T. Lancaster and S. J. Blundell, *Quantum Field Theory for the Gifted Amateur*. Oxford University Press, 2014.

9

The path of extremal action

After going through many strange and unusual things specific to quantum mechanics, now will be a good time to return to the foundations and reconsider its similarity and differences from classical mechanics. We will look at this from a different angle than before, which should give you a broader perspective of quantum and classical mechanics. Besides, this approach is beautiful – possibly the most beautiful thing in all of theoretical physics, but this is a matter of taste.

Spotlight: Richard Feynmann

Richard Feynman (1918–1988) was an American theoretical physicist. He was awarded the Nobel Prize in Physics (in 1965, together with Shin-Itiro Tomonaga and Julian Schwinger) for the development of quantum electrodynamics. Feynman participated in the Manhattan Project, developed the path integral approach to quantum mechanics, and obtained a number of other important results. In particular, he finished the theory of superfluid helium built by Lev Landau, and his diagram technique helped elucidate the concept of quasiparticles introduced by Landau.

Feynman and Landau were much alike in many respects. Both had flamboyant personalities, both were brilliant, both achieved great results across the spectrum of theoretical physics, and both left behind great textbooks.

The Feynman Lectures on Physics remains one of the best undergraduate courses on the subject. Characteristically, for each three copies of English editions of this book ever sold, its Russian translation sold two.

Besides that, Feynman published higher-level physics textbooks, popular books on physics and other subjects. He played a key role in finding the reason behind the *Challenger* disaster. Feynman predicted and encouraged the development of nanotechnology and – what is more important for our purpose – quantum computing.

And yes, Feynman played bongo drums. Landau did not.

Variational approach to classical mechanics

The usual approach to classical mechanics is via the three Newton's laws. For example, for a system of N material points – massive particles of zero size – we need to solve the N vector (or $3N$ scalar – one for each Cartesian component) equations of motion

$$m_i \frac{d^2 r_i}{dt^2} = F_i\left(\left\{r, \frac{dr}{dt}\right\}\right)$$

for the positions of the particles. The summary force F_i acting on the ith particle depends on positions and, possibly, velocities of all the particles and on some external agents (like in the case of a gas placed in the gravity field).

This is not necessarily the simplest way of handling the problem, if instead of material points you are given a system of rigid bodies with constraints – as the simplest example assumes that some of the material points are connected by rigid weightless rods or must slide along fixed tracks (Figure 9.1). Of course, we can still write down the set of Newton's equations, but because of the constraints some of the coordinates must be eliminated.

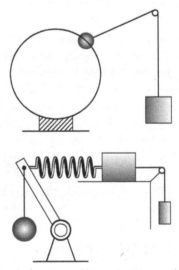

Figure 9.1 A mechanical system with constraints

Even a simple case like a pendulum can cause certain trouble (Figure 9.2). Everything here is in the same plane, so we only need two coordinates of the bob, x and y. They are not independent, since the length of the pendulum,

$$l = \sqrt{x^2(t) + y^2(t)},$$

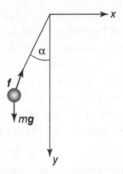

Figure 9.2 A pendulum

is fixed. There are two forces acting on the bob, gravity, mg, and the reaction of the rod, f. The Newton's equations are (it is convenient to direct the y axis downwards):

$$m\frac{d^2x}{dt^2} = f_x = -f(t)\frac{x(t)}{l};$$

$$m\frac{d^2y}{dt^2} = f_y + mg = -f(t)\frac{y(t)}{l} + mg.$$

These equations include the unknown time-dependent force $f(t)$, which we do not need to know anyway: the rod of the pendulum is assumed rigid and unbreakable. On the other hand, in the second equation we can replace the function $y(t)$ with $\sqrt{l^2 - x^2(t)}$. Then we find from the first equation, that $f(t) = -\frac{ml}{x(t)}\frac{d^2x}{dt^2}$, substitute this into the second equation, and obtain quite an ugly equation to solve.

This is, of course, an exaggeration – we know that a pendulum is best treated by the equivalent of Newton's laws for flat rotations, where the convenient coordinate is the angle α. But if we did not know that and had to deal with a more complex system from Figure 9.1, we would most probably waste a lot of time and effort on dealing with unnecessary complications (like getting rid of $f(t)$ and finding the second derivative of $\sqrt{l^2 - x^2(t)}$).

Instead a much more economical approach was developed (primarily by Euler and Lagrange). One is tempted to think that its development was spurred on by the demands of fast growing industry, with the machinery presenting exactly the kind of complex mechanical systems that are hard to treat based on Newton's equations. Ironically, while the industry grew the fastest in Britain, Euler and Lagrange lived and worked on the continent. Maybe this was a matter of their different philosophies.

The original Newtonian approach to mechanics is based on the cause-and-effect chain of events. The force produces the acceleration; therefore both the velocity and the position changes; and this changes the force, and so on. The chain is started by some initial cause – the initial conditions (positions and velocities at the starting moment, which determine all the consequent evolution of the system).

The variational approach is different. Here instead of looking at the motion locally, we consider it globally, in its wholeness.

We are now revisiting some of the notions, which we have touched upon in Chapters 1 and 2.

First, we introduce a set of generalized coordinates, $\{q\}$, and velocities, $\{\dot{q}\} = \left\{\frac{dq}{dt}\right\}$. These coordinates should fully describe the position of our system, but otherwise they can be anything. For a pendulum we only need a single coordinate, the angle α between the rod and the vertical. For a more complex contraption we could use positions of certain masses, extensions of springs, angles between levers, and so on. These are the coordinates in the generalized configuration space of our system.

Then we express the kinetic energy of the system, $T(\{\dot{q}\})$, and its potential energy, $U(\{q\})$, in terms of these coordinates and velocities. (We consider here the simplest (and the most common in mechanics) case, when there are no terms, which would depend on both q and $\dot{q}$.)

Then we introduce the Lagrangian function, or simply Lagrangian:

$$L(\{q, \dot{q}\}) = T(\{\dot{q}\}) - U(\{q\})$$

and define the *action* as its integral along a given path $q_j(t)$, $j = 1\ldots N$, which the system follows between the initial and final times:

$$S[q(t)] = \int\limits_{t_i}^{t_f} dt\, L(\{q, \dot{q}\}).$$

This is the general definition of action. Note that it depends on the whole function $q(t)$ – a path in the multidimensional configuration space. (A usual function depends only on a single *point* – a moment of time, a position, etc.) Therefore action is called a *functional* and is written as $S[q(t)]$, with square brackets – to better distinguish it from a function of a function (like $\exp(\cos(t))$).

Now we can formulate the *variational principle of classical mechanics (alias the principle of extremal (or stationary) action)*:

Every physical motion realizes the extremum (a minimum or a maximum) of the action along its path:

$$\delta S = 0.$$

This means the following. Suppose the actual motion of a system between its initial configuration q_i at the initial moment t_i and its final configuration q_f at the moment t_f is described by the function $\overline{q}(t)$.[1] The action on this path is $\overline{S} = S[\overline{q}(t)] = \int_{t_i}^{t_f} dt\, L(\overline{q}, \dot{\overline{q}})$.

Then, if we slightly shift the path, that is, replace $\overline{q}(t)$ with $\overline{q}(t) + \delta q(t)$ (an infinitesimally small function $\delta q(t)$, which is zero at $t = t_i$ and $t = t_f$, is called a *variation*) (Figure 9.3), the action will not change:

$$\delta S[\overline{q}(t)] = S[\overline{q}(t) + \delta q(t)] - S[\overline{q}(t)] = 0.$$

To be more precise, if $\overline{q}(t)$ is replaced with $\overline{q}(t) + \varepsilon \cdot \eta(t)$, where $\epsilon \ll 1$ is a small number, and $\eta(t)$ is an arbitrary function, which is zero at $t = t_i$ and $t = t_f$, then the action will become

---

1 I write 'configuration' to stress that the set of generalized coordinates q is not necessarily a set of positions in space. They determine a point in the generalized configuration space.

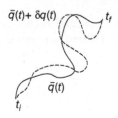

Figure 9.3 Variation of a path

$\bar{S} + \epsilon^2 \cdot S_1$. As $\epsilon \to 0$, the correction to the action will vanish faster than the change to the path. This is like what we have when finding a minimum or a maximum of a usual function.

Key idea: Taylor expansion near an extremum

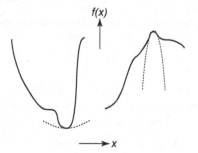

Let a function $f(x)$ have an extremum (a minimum or a maximum) at some point $\bar{x}$. Then $\frac{df}{dx} = 0$ at $x = \bar{x}$, and the Taylor expansion at this point is

$$f(x) = f(\bar{x}) + \frac{1}{2}\frac{d^2 f}{dx^2}(x - \bar{x})^2 + \ldots$$

and the correction to the value of the function is proportional to the *square* of the deviation from the point of extremum.

Key idea: Newtonian and Lagrangian differences

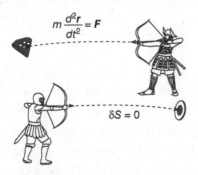

$$m\frac{d^2r}{dt^2} = F$$

$\delta S = 0$

The difference between the Newtonian and the Lagrangian
formulations of mechanics can be compared to the difference
between the Eastern and Western approaches to archery. In the
West, you want to hit the target, and subject all your actions to this
goal. In the East, you want to correctly follow the right procedure
step by step – and if you do, you will hit the target.

The variational approach reformulates all of mechanics in terms
of 'final causes': instead of looking at the evolution step by step,
the system is interested in the final outcome and shapes its path
correspondingly. In the 18th century, Pierre-Louis Moreau de
Maupertuis formulated the principle in its modern form as the
principle of least action. In its time this caused much philosophical
stir, since the notion of a physical system 'knowing' its final goal
seemed to have theological implications. On the second thought,
though, this 'foreknowledge' is nothing else than the Laplacian
determinism of the Newtonian dynamics (Chapter 1).

The practical usefulness of the variational approach is not in
its literal application. One does not calculate the action for all
possible paths to choose the extremal one. But starting from this
principle one can derive local equations, which are usually much
simpler than the original Newtonian ones.

There is a whole branch of mathematics, *calculus of variations*,
pioneered by Euler, which is devoted to this sort of problem.

In particular, from the variational principle the Lagrange, or Euler–Lagrange, equations can be derived:

$$\frac{d}{dt}\left(\frac{\partial L}{\partial \dot{q}_j}\right) - \frac{\partial L}{\partial q_j} = 0.$$

These equations should be written for every generalized coordinate. The resulting set is equivalent to the Newtonian equations, but is much more convenient to deal with.

Take, for example, the pendulum. Its state is completely described by the angle α. The potential energy is $U(\alpha) = mgl(1 - \cos\alpha)$ and the kinetic energy $T(\dot{\alpha}) = m(l\dot{\alpha})^2/2$. Then the Lagrangian is

$$L(\alpha, \dot{\alpha}) = \frac{m(l\dot{\alpha})^2}{2} - mgl(1 - \cos\alpha).$$

The Lagrange equation, $\frac{d}{dt}(ml^2\dot{\alpha}) + mgl\sin\alpha = 0$, or $\frac{d^2\alpha}{dt^2} = -\frac{g}{l}\sin\alpha$, is the usual equation for a pendulum, which could be obtained without using the Lagrange approach. Still, in this way it is obtained effortlessly, and the advantages of the method become evident for more complex systems.

Practically all of theoretical physics (including electrodynamics, special and general relativity, hydrodynamics, etc.) can be presented starting from variational principles. For example, Landau and Lifshitz take this approach in their multivolume *Course of Theoretical Physics*, which helped educate generations of physicists worldwide.

A reasonably hard exercise

Try to apply the Lagrange method to one of the mechanisms shown in Figure 9.1.

1 Choose convenient generalized coordinates.
2 Write the potential and kinetic energy in terms of these coordinates and generalized velocities.
3 Write the Lagrange function.
4 Write the Lagrange equations.
5 Feel really good about it.

Variational approach to quantum mechanics

Let us now return to the literal interpretation of the variational principle. Suppose we could calculate action for all possible paths, $S[q(t)]$, compare them and choose the extremal one. How could we simplify this task?

To begin with, we can calculate for each path $q_j(t)$ a fast oscillating function, for example, $\cos(\lambda S[q_j(t)])$, where λ is a large number. A small change in $q(t)$ will produce a small change in $S[q(t)]$, but if the factor λ is large enough, the cosine can change all the way between minus one and one. Now we simply add together all the contributions from different paths. In the resulting sum, $\sum_j \cos(\lambda S[q_j(t)])$, the contributions from different paths will mostly cancel out (see Figure 9.4): for each path $q_j(t)$ there will be a nearby path $q_j(t) + \varepsilon \cdot \eta(t)$ (the small number ε is of the order of $1/\lambda$) such that

$$\cos(\lambda S[q_j(t) + \varepsilon \cdot \eta(t)]) \approx \cos(\lambda S[q_j(t)] + \pi) = -\cos(\lambda S[q_j(t)]).$$

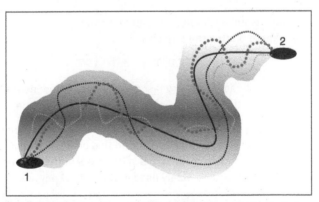

Figure 9.4 Contributions from different paths. The solid line is the surviving extremal path, which corresponds to the classical motion.

The exception will be such path $\overline{q}(t)$, that its variation $\overline{q}(t) \rightarrow \overline{q}(t) + \varepsilon \cdot \eta(t)$ only affects the action in the second order in ε:

$$\overline{S} \rightarrow \overline{S} + \varepsilon^2 S_1,$$

and the arguments of the cosine will change by the amount of the order of $\epsilon^2 \lambda \cdot \pi \sim \epsilon \cdot \pi$, and not $\sim \pi$. Therefore the contributions to the sum from all the paths close to $\bar{q}(t)$ will have the same sign and, instead of cancelling each other, will add up. The more paths we include in our sum, the more dominant will be this contribution. As a result, the one path that eventually survives will be precisely the path $\bar{q}(t)$, for which the action takes the extremal value. As we have said, it can be obtained from the Lagrange equations, that is, from classical mechanics.[2]

Key idea: Extremal path

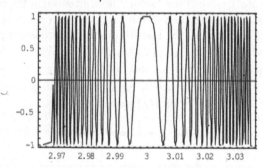

To illustrate the survival of the contribution from the extremal path, here I have plotted the function

$$\cos(100000(x-3)^2).$$

As you can see, it oscillates very fast everywhere, except near the point $x = 3$, where the argument of the cosine has a minimum.

If we wanted to approximately calculate the integral of this function, we could consider only the interval around this point – the rest would almost completely cancel. This is the basis of the well-known *method of stationary phase*, widely used in theoretical physics.

Our task here is more complex – we must 'integrate over paths', but the same idea is used if we want to approximately evaluate a path integral.

---

2 Similar reasoning explains why light, being a wave, propagates along straight lines. There the role of the small parameter plays the light wavelength λ: the effects of light interference and diffraction appear on the scale comparable to λ.

But now we can turn around and say: if we *start* our reasoning from the sum like $\sum_j \cos(\lambda S[q_j(t)])$ and consider which term contributes the most, we will inevitably arrive in the end at the principle of stationary action. The principle, which looked like some mysterious revelation from the top and caused so many philosophical speculations, becomes simply a mathematical necessity.

The question of what is the large number λ is easily answered: it is $\frac{1}{\hbar}$. Indeed, then the argument of the oscillating function is a dimensionless number, $\frac{S}{\hbar} = 2\pi \frac{S}{h}$, that is, 2π times the action of physical system in the units of action quantum. An extra quantum makes the oscillating function run the full circle.

This was the inspiration for the approach suggested by Dirac in 1932 and realized by Feynman a quarter of a century later.

The oscillating function used is $\exp(iS[q_j(t)]/\hbar) = \cos(S[q_j(t)]/\hbar) + i\sin(S[q_j(t)]/\hbar)$. The question is: how do we calculate the contributions from different paths, and contributions to *what?*

Feynman's path integrals

Consider a quantum system moving from the point 1 (that is, (q_i, t_i)) to the point 2 (q_f, t_f) (Figure 9.4) in the configuration space. In other words, the system was at the initial moment of time in the state $|\Psi_i\rangle = |q_i, t_i\rangle$, and at the final moment in the state $|\Psi_f\rangle = |q_f, t_f\rangle$. We will calculate the scalar product of these two states, $\langle \Psi_i | \Psi_f \rangle$, which is also called the *transition amplitude*.

This is a very reasonable quantity to calculate. It gives us a quantitative measure of how different the final state is from the initial one. The probability to arrive at the given final state from the given initial state (that is, *transition probability*) is the square modulus of the transition amplitude.

What Feynman succeeded in doing was to demonstrate the following:

The transition amplitude $\langle \Psi_i | \Psi_f \rangle$ can be expressed as the *path integral* of the exponent $\exp\left(\frac{i}{\hbar} S[q(t)]\right)$ over all possible paths, which connect the initial configuration of a quantum system, (q_i, t_i), with its final configuration, (q_f, t_f).

To have a feeling of what this 'path integral' is, we consider a system, which consists of a single particle of mass m, which moves along the x axis, so that the generalized coordinate q is simply its position x. The Lagrangian of the particle is

$$L(q, \dot{q}) = \frac{m}{2} \left(\frac{dq}{dt} \right)^2 - V(q),$$

where $V(q)$ is some external potential.

In order to calculate the path integral, we imitate the approach towards calculation of ordinary integrals. There, the integration interval was sliced in such small pieces, that the integrated function in each slice is approximately constant, and then the contributions from each slice are added.

Here we will do the same: slice the interval $[t_i, t_f]$ in N pieces of length Δt (Figure 9.5). A given path $q(t)$ will be sliced into pieces as well. The particle at the moment $t_1 = t_i + \Delta t$ will have the coordinate q_1, at $t_2 = t_i + 2\Delta t$ the coordinate q_2, and so on. We will approximate the Lagrangian on the interval $[q_{j-1}, q_j]$ with

$$L_j = \frac{m}{2} \left(\frac{q_j - q_{j-1}}{\Delta t} \right)^2 - V\left(\frac{q_j + q_{j-1}}{2} \right).$$

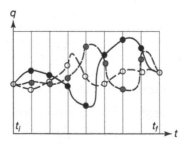

Figure 9.5 Slicing the interval and trajectories

The action integral is approximated by a sum:

$$S[q(t)] = \int_{t_i}^{t_f} L(q,\dot{q})\, dt \approx \sum_{j=1}^{N-1} L_j\, \Delta t.$$

We will introduce this approximation in the exponent:

$$\exp\left(\frac{i}{\hbar} S[q(t)]\right) \approx \exp\left(\frac{i}{\hbar} \sum_{j=1}^{N-1} L_j\, \Delta t\right).$$

This is an approximate expression, corresponding to one sliced path. But it allows us to take into account *all* such paths. Indeed, if we integrate this expression over all intermediate positions from minus to plus infinity, we include all possible paths running between (q_i, t_i) and (q_f, t_f):

$$\text{'path intergral'} \approx \int_{-\infty}^{\infty} dq_1 \int_{-\infty}^{\infty} dq_2 \dots \int_{-\infty}^{\infty} dq_{N-1} \exp\left(\frac{i}{\hbar} \sum_{j=1}^{N-1} L_j \Delta t\right).$$

If we now take $\Delta t \to 0$ (respectively, $N \to \infty$), we will have an infinite number of integrations over the q's, but we will restore the exact, and not approximate, action along each of these paths, $S[q(t)] = \int L(q,\dot{q})$ As a more accurate analysis demonstrates, one should include certain normalization factors $C/\sqrt{\Delta t}$ in each of the integrals over q_j before taking the limit $\Delta t \to 0$, so this is not such an easy operation as a transition from a sum to an ordinary integral. Nevertheless it is quite doable. The result is

$$\text{'path intergal'} = \lim_{\substack{\Delta t \to \infty \\ N \to \infty}} \int_{-\infty}^{\infty} \frac{C dq_1}{\sqrt{\Delta t}} \int_{-\infty}^{\infty} \frac{C dq_2}{\sqrt{\Delta t}} \dots \int_{-\infty}^{\infty} \frac{C dq_{N-1}}{\sqrt{\Delta t}} \exp\left(\frac{i}{\hbar} \sum_{j=1}^{N-1} L_j \Delta t\right)$$

$$= \int_{q_i(t_i)}^{q_f(t_f)} Dq \exp\left(\frac{i}{\hbar} S[q(t)]\right)$$

The capital D in Dq stresses that the path integral is not any garden variety integral. It is a much more complex mathematical object called a *functional integral*, but, as I have said, it can be calculated, at least approximately or numerically, when necessary. Note that all these integrals only contain usual functions, no operators.

So, the transition amplitude between the initial and final configurations of a quantum system can be written as:

$$\langle \Psi_i | \Psi_f \rangle = \int_{q_i(t_i)}^{q_f(t_f)} Dq \, \exp\left\{ \frac{i}{\hbar} \int_{t_i}^{t_f} L(q, \dot{q}) \, dt \right\}.$$

One can derive from the path integral formulation all the equations of quantum mechanics we have encountered before, even in the case of many-body systems. In the latter case we can define Feynman integrals over *fields*. Instead of the Lagrangian we will have the Lagrangian *density* $\mathcal{L}$ (which will be integrated over all of space to obtain the Lagrangian) and, instead of Dq, we may encounter $D\bar{\psi} D\psi$ (pointing at the integration over all possible configurations of fields).

We know that quantum fields can be bosonic or fermionic. With bosons there is no problem – the corresponding fields in path integrals are usual functions, with real or, at worst, complex values. With fermions we would seem to get into trouble; there are no 'usual numbers', which would *anti*commute. Fortunately, Felix Berezin discovered the method of calculating it using special numbers (so-called Grassmann variables). We will not discuss how it works (though this is a fascinating and rather simple topic), but will just assure you that it does work.

Besides the 'philosophical' attractiveness (the explanation of where the variational principles of physics are coming from), there are other advantages of using the path integral formulation. The main one is that the path integrals include the Lagrangian, and not the Hamiltonian.

This is not an important point as long as we are happy with the non-relativistic world, where space and time do not mix.

The Schrödinger equation, the Heisenberg equation, the von Neumann equation – everywhere time is a variable apart. For example, the Schrödinger equation contains the first order time derivative, and the second order position derivative.

The problems appear when one can no longer neglect the relativistic effects. As we have known since 1907, courtesy of Hermann Minkowski, what seems to us 'three-dimensional space + one-dimensional time' is actually the 'four-dimensional spacetime'. This means, in particular, that coordinates and time get mixed, if we deal with objects moving with relativistic speeds (that is, with speeds comparable to the speed of light in vacuum). Therefore a quantum theory of such objects can no longer treat them unequally, like in the Schrödinger equation. It must be what is called *Lorentz invariant*.

Spotlight: Schrödinger and Dirac equations

Schrödinger first tried to derive a Lorentz invariant equation for an electron, but it did not yield the expected energy levels of a hydrogen atom known from the experiment. Then Schrödinger decided to try a non-relativistic approximation and fully succeeded. Later it turned out that his first attempt was actually a correct equation, but for a different kind of particle: a relativistic boson instead of a relativistic fermion.

Dirac derived a correct relativistic equation for electrons in 1928. Analysing it, he came to the concept of *antiparticles* and predicted the existence of antielectrons (i.e. positrons), which were discovered by Carl Anderson in 1932.

This is where the Lagrangian-based approach is especially handy. For quantum fields, the action is an integral of the Lagrangian density over both space and time, and it can be rewritten as an integral over the 'hypervolume' of some properly selected piece of the Minkowski spacetime.

$$S[\bar{\psi},\psi] = \int_{t_i}^{t_f} dt \iiint d^3r \mathcal{L}\big(\bar{\psi}(r,t),\psi(r,t)\big) = \int d^4X \mathcal{L}\big(\bar{\psi}(X),\psi(X)\big).$$

Here we denote the set of four spacetime coordinates by X. Now the path integral

$$\int D\bar{\psi}\, D\psi\, e^{\frac{i}{\hbar}\int d^4 X \mathcal{L}(\bar{\psi},\psi)}$$

does *not* make any explicit difference between space and time, and it can be used as the basis for a relativistic quantum field theory. Of course, you still do not know how to calculate such an integral, but then, we are only learning here *about* quantum theory. If you decide to study quantum theory at a rather advanced level, you will know how.

Charged quantum particles in an electromagnetic field, path integrals, the Aharonov–Bohm effect and the Berry phase*

The path integral formulation of quantum mechanics is practically useful in usual, non-relativistic quantum mechanics as well. Take, for example, a very interesting effect, which was predicted theoretically by Yakir Aharonov and David Bohm in 1959 and confirmed by experiments soon after that. Remarkably, this prediction could be made back in the 1920s, as soon as Schrödinger proposed his equation. On the other hand, such a delay is minor compared to the approximate 1500 years that elapsed between the Hellenic times (when there existed both the theory and the technology required to build a telescope and a microscope) and Galileo's time (when it was actually invented). By the way, the experimental confirmation of the Aharonov–Bohm effect was obtained soon after it was predicted.

> The essence of the Aharonov–Bohm effect is that an electromagnetic field can act on a quantum particle even when there is no electromagnetic field at the particle's location.

Case study: Vector potential

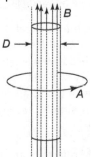

We know that the electrostatic field **E** can be described by the scalar potential ϕ (alias voltage). It is called *scalar* potential because it is determined in each point by a single number (voltage).

The magnetic field **B** can be similarly described by a *vector potential* **A**. The vector potential is related to the magnetic field the following way. The *magnetic flux* Φ through a surface with an area S is – roughly speaking – the number of the force lines of the magnetic field penetrating this surface. (If the magnetic field **B** is uniform and orthogonal to S, then the magnetic flux is simply Φ = BS). Then the vector potential is so chosen that its integral along the boundary Γ of this area equals Φ.

Mathematically this is expressed as

$$\iint_S \boldsymbol{B} \cdot d\boldsymbol{s} = \oint_\Gamma \boldsymbol{A} \cdot d\boldsymbol{l}.$$

For example, take a solenoid (a long cylindrical coil of diameter D and length L (L ≫ D), through which flows the electric current I). It is known from elementary electromagnetic theory, that the magnetic field of a solenoid *inside* it is uniform and proportional to NI, where N is the number of turns of the wire. Near the ends of the solenoid this is no longer true, but if the solenoid is really long, this can be neglected. Remarkably, if the solenoid is infinitely long, then there is *no* magnetic field outside. If it is finite, but long, the magnetic field will appear on the outside, but only far away from the solenoid. Therefore, as long as we keep close enough, we can still consider the solenoid as infinitely long.

The magnetic flux through the solenoid is $\Phi = B \cdot \frac{\pi D^2}{4}$. Since there is no magnetic field outside, then the same magnetic flux penetrates *any* surface, which is crossed by the solenoid. We can then choose the vector potential in a very symmetric way: let it go along the circles, centred on the axis of the solenoid. Then the integral of **A** along a circle of radius R is simply the length of the circle times the modulus of **A:**

$\oint \boldsymbol{A} \cdot d\boldsymbol{l} = 2\pi R \, A(R) = \Phi$. Therefore the modulus of the vector potential of a solenoid is $A(R) = \frac{BD^2}{8R}$.

Of course, the effect can be obtained directly by solving the Schrödinger equation, but the path integral approach allows an especially straightforward presentation.

To begin with, we must figure out how to take into account the interaction of a quantum particle with the electromagnetic field. With the electric field there was no problem: we added the electrostatic energy to the potential energy term (like the repulsive Coulomb potential in the problem of α-decay). The magnetic field is included through its vector potential, which enters the expression for the *kinetic* energy. To be more specific:

In classical physics, in order to take into account the interaction of a particle with the electric charge Q with the magnetic field, its momentum p in the Hamiltonian must be replaced by $p - \frac{Q}{c} A$, where A is the vector potential of the magnetic field. (In quantum physics, the momentum operator $\hat{p}$ in the Hamiltonian operator must be replaced by $\hat{p} - \frac{Q}{c} A$.)

We are still using the Gaussian units here, this explains the factor $\frac{1}{c}$, with c being the speed of light. The justification of such an operation – at the level we care about here – is that the electromagnetic field has its own momentum, and this is the simplest (and correct) way to account for it.

But how do we include the magnetic field in the Lagrangian function? Recall that while the Hamiltonian is the *sum* of kinetic and potential energy terms, the Lagrangian function is their *difference*: $H = T + U$, $L = T - U$. Therefore all we need is to rewrite the kinetic energy term through velocities rather than through momenta.

Take a single charged particle of mass M. Then its kinetic energy in the magnetic field is

$$T = \frac{\left(p - \frac{Q}{c}A\right)^2}{2M} = \frac{p^2}{2M} - \frac{Q}{Mc}A \cdot p + \frac{Q^2}{2Mc^2}A^2.$$

The last term relates to the energy of the magnetic field itself and is of no concern to us. The first term is just the kinetic energy of the particle *without* any magnetic field. All the particle-field interaction is contained in the second term, which we can rewrite as

$$-\frac{Q}{c}A \cdot v,$$

where $v = \dot{x} = \frac{p}{M}$ is the velocity of the particle. Therefore in the presence of the magnetic field the Lagrangian of a charged particle becomes

$$L(x,\dot{x},A) = L(x,\dot{x}; A = 0) - \frac{Q}{c}A \cdot \dot{x} = L(x,\dot{x}; A = 0) - \frac{Q}{c}A \cdot \frac{dx}{dt}.$$

Now let an electron propagate between the points A (e.g. the source of electrons – a cathode) and B (e.g. a screen, sensitive to electrons) past a solenoid with the magnetic flux Φ. We make the solenoid walls impenetrable to electrons. We also enclose our contraption in an electron-impenetrable box (for reasonable electron energies a wooden box would do), with the end of the solenoid sticking far, far out – this guarantees that where there are electrons, there is no magnetic field, and where is the magnetic field, there are no electrons.

Let us calculate the probability amplitude for an electron to go from A to B:

$$\langle A|B\rangle = \int\limits_{A,t_A}^{B,t_B} Dx \exp\left\{\frac{i}{\hbar}\int\limits_{t_A}^{t_B} L(x,\dot{x})dt\right\}.$$

The square modulus of this quantity gives the probability for an electron to end up in the point B on the screen. It therefore describes the interference pattern, which we shall see after sending a number of electrons, one by one, past the solenoid.

Substituting in the Lagrangian the term with the vector potential, we will see that the action exponent is

$$\exp\left\{\frac{i}{\hbar}\int\limits_{t_A}^{t_B} L(x,\dot{x},A)\,dt\right\} = \exp\left\{\frac{i}{\hbar}\int\limits_{t_A}^{t_B} L_0(x,\dot{x};A=0)\,dt\right\}\exp\left\{\frac{i}{\hbar}\int\limits_{t_A}^{t_B}\frac{e}{c}A\cdot\frac{dx}{dt}dt\right\}$$

$$= \exp\left\{\frac{i}{\hbar}\int\limits_{t_A}^{t_B} L(x,\dot{x};A=0)\,dt\right\}\exp\left\{\frac{ie}{\hbar c}\int\limits_{A}^{B} A\cdot dx\right\}.$$

When calculating this path integral, we must take into account all the possible paths a particle can take between A and B, which satisfy the physical constraints, e.g. do not cross the solenoid and do not leave the wooden box. Take a pair of such paths, of which one goes to the left, and another to the right of the solenoid (Figure 9.6). Their contribution to the transition amplitude is, say,

$$C_1 \exp\left\{\frac{i}{\hbar}S_{path1}\right\}\exp\left\{\frac{ie}{\hbar c}\int_{path1} A\cdot dx\right\}$$

$$+ C_2\exp\left\{\frac{i}{\hbar}S_{path2}\right\}\exp\left\{\frac{ie}{\hbar c}\int_{path2} A\cdot dx\right\}.$$

Here S_{path1} and S_{path2} are actions in the absence of the magnetic field, and we choose for simplicity a pair with $S_{path1} = S_{path2} = S$.

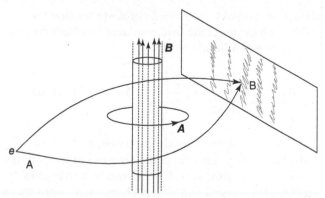

Figure 9.6 The Aharonov–Bohm effect: a series of electrons flying past a solenoid. The interference pattern on the screen will change if the magnetic field in the solenoid changes, even though the electrons cannot appear in the regions where the magnetic field is nonzero

The contribution of this pair of paths to the *probability* for the electron to get from A to B is the square modulus of the above expression. When calculating it, we will obtain several terms, of which this one is of interest:

$$C_1 \exp\left\{\frac{i}{\hbar}S\right\}\exp\left\{\frac{ie}{\hbar c}\int_{path\,1} A \cdot dx\right\} \times C_2 \exp\left\{-\frac{i}{\hbar}S\right\}\exp\left\{-\frac{ie}{\hbar c}\int_{path\,2} A \cdot dx\right\}$$

$$+ C_1 \exp\left\{-\frac{i}{\hbar}S\right\}\exp\left\{-\frac{ie}{\hbar c}\int_{path\,1} A \cdot dx\right\}$$

$$+ C_2 \exp\left\{\frac{i}{\hbar}S\right\}\exp\left\{\frac{ie}{\hbar c}\int_{path\,2} A \cdot dx\right\}$$

$$= C_1 C_2 \left[\exp\left\{\frac{ie}{\hbar c}\int_{path\,1} A \cdot dx - \frac{ie}{\hbar c}\int_{path\,2} A \cdot dx\right\}\right.$$

$$\left. + \exp\left\{-\frac{ie}{\hbar c}\int_{path\,1} A \cdot dx + \frac{ie}{\hbar c}\int_{path\,2} A \cdot dx\right\}\right].$$

Recalling that $\exp(ia) + \exp(-ia) = 2 \cos a$, we see that the dependence on the magnetic field produces oscillations in the probability of going from A to B:

$$\left| \langle A | B \rangle \right|^2 \sim \sum_{path1, path2} \cos \left\{ \frac{e}{\hbar c} \left(\int_{path\,1} A \cdot dx - \int_{path\,2} A \cdot dx \right) \right\}.$$

This does not seem a very useful expression, with all its integrals. But let us reverse the path 2: instead of integrating the vector potential from A to B we integrate it from B to A (Figure 9.6). The integral will simply change sign (since we go along the same path, but in the opposite direction). But then the sum of two integrals becomes one integral along the *closed* path from A to B and back to A, which goes around the solenoid:

$$\frac{e}{\hbar c} \left(\int_{path\,1} A \cdot dx - \int_{path\,2} A \cdot dx \right) = \frac{e}{\hbar c} \oint A \cdot dx = \frac{e}{\hbar c} \Phi.$$

On the last step we have used the defining property of the vector potential. Now we see that each pair of paths with $S_{path1} = S_{path2} = S$ will add to the probability $\left| \langle A | B \rangle \right|^2$ a term that contains *the same* oscillating factor $\cos \frac{e\Phi}{\hbar c}$. There will be also contributions from other terms, but they will not completely erase these oscillations. Therefore the interference pattern on the screen will change, if the magnetic field in the solenoid changes – even though the electrons never go into the regions of space, where the magnetic field B is nonzero! The vector potential of the field is nonzero there and this is enough, despite the fact that there is *no* force acting on the electron. A classical charged particle would not notice anything.

This beautiful and counterintuitive result is purely quantum mechanical, and it was observed in a number of experiments. For example, if you measure the electrical conductance of a small (micron-sized) metal ring, it will oscillate as you change the magnetic field penetrating the ring (conductance is, after all, determined by the probability that an electron can get from the one electrode to another).

The oscillating factor $\cos\frac{e\Phi}{hc}$. can be rewritten as $\cos 2\pi\ \Phi/\tilde{\Phi}_0$, to stress that the period of oscillations is

$$\tilde{\Phi}_0 = \frac{hc}{e}.$$

This is so-called *normal flux quantum* (called 'normal' to distinguish it from the *superconducting* flux quantum, which is two times smaller and will be discussed in Chapter 10). If you add to the magnetic flux in our solenoid any integer number of flux quanta, nothing changes.

The normal flux quantum is a small magnetic flux indeed. In the usual units (where the magnetic field is measured in teslas, current in amperes, charge in coulombs and the magnetic flux in *webers*: 1 Wb = 1 T × 1 m²):

$$\tilde{\Phi}_0 = \frac{h}{e} = 1.03...\times 10^{-15}\,\text{Wb}.$$

The Aharonov–Bohm effect is an example (the simplest one) of the so-called *Berry phase* (after Michael Berry). The Berry phase is φ in the phase factor $\exp(i\varphi)$, which the wave function of a particle can acquire after going around a closed loop (generally in the Hilbert space), when naively nothing should have changed:

$$\left|\Psi\right\rangle \to \left|\Psi\right\rangle e^{i\varphi}.$$

Of course, the phase factor does not affect the state vector itself – after all, the probability is given by its modulus squared. But if there is interference between two vectors with different Berry phases, as in the case of the Aharonov–Bohm effect, there can be observable consequences.

Now we can ask whether such effects could happen to large systems. Suppose a *big* particle can go from A to B along two mirror-symmetric ways around the solenoid, with *exactly the same* actions S, and the only difference due to the vector potential. Should not we see the Aharonov–Bohm effect? The particle can be big – even though the action S is much, much greater, than $\hbar$, it cancels out. The same reasoning should apply to a big object going through a double slit. Nevertheless we do not observe *big* objects doing these kinds of things.

In the early and not so early days of quantum theory there was the intuition that big objects are just prohibited from doing this. Nevertheless the past 20 or so years of research have shown that this is not necessarily so, and that the problem of quantum-classical transition (that is, how the cosy and familiar laws of classical physics emerge from the strange quantum rules) requires a thorough investigation. It was especially spurred on by the quest for quantum computing, so we will wait until Chapter 12 to discuss this matter.

Fact-check

1 In classical mechanics, a physical motion realizes
  a a maximum of the Lagrangian
  b a minimum of the Lagrangian
  c an extremum of the action
  d a minimum of energy

2 The Euler–Lagrange equations of motion
  a follow from the variational principle
  b are equivalent to the Newtonian equations
  c only apply to mechanical systems
  d determine the trajectory to the lowest approximation

3 A Feynman's path integral determines
  a the extremal path between the initial and final quantum states
  b the transition amplitude between the initial and final quantum states
  c the transition probability between the initial and final quantum states
  d the superposition of the initial and final quantum states

4 The path integral approach relies upon
  a the quantum superposition principle
  b the momentum conservation law
  c the Pauli exclusion principle
  d the principle of the least action

5 The Aharonov–Bohm effect demonstrates that
  a the phase of a charged quantum particle depends on the magnetic field
  b a charged quantum particle interacts with the magnetic field
  c a charged quantum particle interacts with the vector potential
  d a quantum particle can travel through a solenoid

6 The action $S[\overline{q}(t)]$ is extremal. This means that
  a $S[\overline{q}(t)]$ does not change with time
  b $S[\overline{q}(t)]$ is the greatest or the smallest possible
  c $S[\overline{q}(t)]$ does not change if $\overline{q}$ is slightly changed
  d $S[\overline{q}(t)]$ can have only one value

7 The Lagrangian of a free particle is $L = \frac{m\dot{q}^2}{2}$. Its Euler–Lagrange equation is
  a $m\ddot{q} - m\dot{q}q = 0$
  b $m\ddot{q} = 0$
  c $m\ddot{q}q = 0$
  d $m\ddot{q} = 0$

8 The path integral approach to relativistic quantum systems is convenient, because
  a it does not contain operators
  b it does not treat time and space coordinates differently
  c it is a classical approximation
  d it is easy to make Lorentz invariant

9 The condition $S \gg \hbar$ means that
  a all paths will contribute to the path integral
  b only extremal paths will contribute to the path integral
  c contributions from different paths will not interfere
  d only contributions from nearby paths will interfere

10 Path integral approach can be applied to quantum fields
  a only if they are relativistic
  b only if they are bosonic
  c only if they are fermionic
  d always

Dig deeper

An instructive introduction to variational approaches is given – as usual – by

R. Feynman, R. Leighton and M. Sands, *Feynman Lectures on Physics*. Basic Books, 2010 (there are earlier editions). See Vol. II (Chapter 19: Principle of least action).

See also Lecture 6 in

L. Susskind and G. Hrabovsky, *The Theoretical Minimum: What You Need to Know to Start Doing Physics*. Basic Books, 2014.

The method of path integrals is best learnt directly from the master:

R. P. Feynman and A. R. Hibbs, *Quantum Mechanics and Path Integrals*. McGraw-Hill Companies, 1965 (there are later editions).

Order! Order!

Quantum systems are – in principle – more complex than the classical ones, in particular those quantum systems that contain many quantum particles. They possess specific, strange and outright weird properties, such as *entanglement*, and their behaviour is, in principle, impossible to model in all detail using classical computers (the fact pointed out by Feynman, which initiated the search for quantum computers). Nevertheless, as we have seen in Chapter 8, there are situations when an approximate treatment of quantum many-body systems are not only possible but also pretty straightforward and accurate. Fortunately, these situations include a large number of scientifically interesting and practically very important cases. We will start with the workhorse of class demonstration experiments: magnetism.

Classical magnets, second order phase transitions and the order parameter

Magnetism is one of the oldest subjects of scientific investigation. In the classic treatise *On the Magnet, Magnetic Bodies, and the Great Magnet of the Earth* published in 1600, William Hilbert (1544–1603) had already stated that magnetic attraction is a separate phenomenon, not to be mixed up with the force exerted by an electrified piece of amber. Since then, the investigation of magnetism became one of the most important areas of physics. Many results, methods and approaches migrated from there to far removed areas of physics, to cosmology, biology and even economics.

A ferromagnet – what is colloquially called a 'magnet' – in equilibrium has a magnetic moment M, which disappears above certain temperature T_C (the Curie point). This is because certain atoms in this system have microscopic magnetic moments of their own, and their interaction with each other tends to align them (Figure 10.1). At high enough temperatures the thermal fluctuations rattle the system so much that this alignment is destroyed, and the magnet loses its magnetization. If we cool the magnet down, the magnetization is restored. (However we will not go any deeper into some important details here, e.g. that a ferromagnet is usually split into regions with different directions of magnetization (domains) and will consider a single domain.) This process is called *phase transition*.

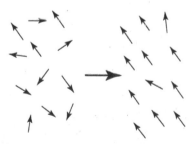

Figure 10.1 Ordering of magnetic moments in a ferromagnet

Classical statistical physics has a number of ways of describing magnets, which successfully reproduce their key features. In a simple theoretical model of a classical magnet we assume that each atom in a crystal lattice has a magnetic moment μs_j (j labels the lattice site, and we neglect all the atoms that do not possess a magnetic moment of their own). Here μ is the absolute value of the atomic magnetic moment (assuming for simplicity that they are all the same), and s_j is a vector of unit length, which shows its direction. The Hamiltonian function (i.e. the energy) of such a magnet is

$$H = -\sum_{jk} J_{jk} s_j \cdot s_k - \mu \sum_j h \cdot s_j.$$

Spotlight: Niels Henrik David Bohr

Niels Henrik David Bohr (1885–1962) was the author of the first quantum theory of atomic structure, and one of the founding fathers of quantum theory. He also received the Nobel Prize in Physics in 1922.

As a young researcher, Bohr received an invitation from Ernest Rutherford to conduct post-doctorate research with him in Manchester (1911). In 1913 he proposed a solution to the puzzles of Rutherford's model of atom and of the atomic spectra (in classical physics atoms would be unstable, atoms of the same elements could have different properties, and their emission spectra should not be anything like what was observed and described by the Balmer's and Rydberg's formulas).

Bohr's *first* major contribution to quantum mechanics was actually made already in 1911, though nobody realized that at the time. In his doctoral thesis, which did not attract much attention, Bohr proved the *Bohr–van Leeuwen theorem* (Hendrika van Leeuwen proved it independently in 1921). The theorem states that a system of classical charged particles in equilibrium in any external electric or magnetic field and at any temperature must have an exactly *zero* magnetic moment.

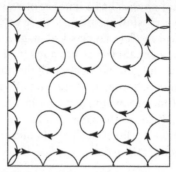

A simple illustration of the Bohr–van Leeuwen theorem. Let us put electrons in a box with ideally reflecting walls and apply a magnetic field. In equilibrium most electrons will run with in circles (say, clockwise) under the action of the Lorentz force, $\boldsymbol{F} = \frac{e}{c}\boldsymbol{v} \times \boldsymbol{B}$; these circular currents will produce a magnetic moment. Nevertheless there will be electrons, which bounce off the walls, move *counter*-clockwise and produce a magnetic moment pointing in the opposite direction. The theorem states that the two contributions always cancel each other *exactly*.

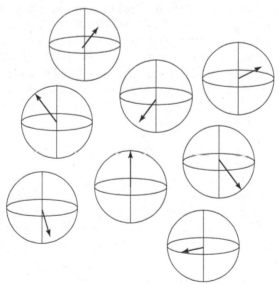

Figure 10.2 Coordinates, which describe a system of N magnetic atoms

Here J_{jk} is the coupling between two magnetic moments (which has the dimension of energy), and h is the external magnetic field. In equilibrium, physical systems have the lowest possible energy.[1] The signs are chosen to ensure that, first, in the presence of the magnetic field each atomic moment tends to align with it, as a compass needle: the second term in the Hamiltonian will correspond to the lowest energy, when for each site $h \cdot s_j > 0$. Second, if all J_{jk} are non-negative, the first term in the Hamiltonian function will give the lowest energy, when all the dot products are positive, $s_j s_k > 0$. This is the case of *ferromagnetic* coupling, as all the magnetic moments will want to point in the same direction. On the contrary, if the J_{jk} were negative, the pairs of coupled moments would want to point in the opposite directions (*antiferromagnetic* coupling).[2] If the couplings can have either sign, especially at random, the situation becomes really interesting and often intractable. But here we are concerned with the ferromagnetic coupling only.

Let us switch off the external magnetic field (put $h = 0$) and look at the average magnetic moment of the system of N magnetic atoms. Here we can use essentially the same approach as in Chapter 1: consider the ensemble of systems in a configuration space. There will be $2N$ coordinates, since the direction of a unit vector s_j is determined by two numbers (for example, the polar angle θ_j changing from zero to π radians, and the azimuthal angle α_j changing from zero to 2π, Figure 10.2).

The ensemble is represented by a 'swarm' of $M \gg 1$ points, each corresponding to a possible state of the system. The average magnetization on the site number j of the Ising magnet is

$\langle s_j \rangle = \lim_{M \to \infty} \frac{1}{M} \sum_{K=1}^{M} s_j^{(k)}$. (The upper index labels the point in the swarm, i.e. a copy of the system in the ensemble.)

It turns out that at high enough temperatures $\langle s_j \rangle = 0$: there is no magnetization. But at the Curie point, $T = T_C$, the same

---

[1] Otherwise it is not really in an equilibrium state. Of course, it can take a *very* long time for a system to reach a true equilibrium. For a glass it can take millennia. For a human, a century is usually quite sufficient.

magnetization appears on each site: $\mu \langle \boldsymbol{s}_j \rangle = M \neq 0,$ and the system as a whole acquires a finite magnetic moment per unit volume. This is called a *second order phase transition*.

Case study: Phase transitions

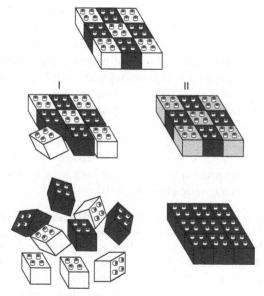

When a substance in equilibrium changes its macroscopic properties as the external conditions (e.g. temperature or pressure) change, we talk about a *phase transition*.

The most common phase transitions are between a solid and a liquid (melting and freezing) and a liquid and a gas (evaporation and condensation). These phase transitions are called the *first order* phase transitions. (The classification is due to Paul Ehrenfest.)

One of their characteristic features is that the first order phase transitions are not 'free of charge': a certain amount of energy in the form of heat must be supplied or taken from the system in the process. For example, it is not enough to bring water to its boiling point – it is necessary to keep the kettle on for a while before all water evaporates, and all this time the temperature of boiling water will stay the same. Similarly, an ice cube in a drink will also stay at

Before the system acquired a magnetic moment, all directions were the same. Once there is magnetization, there is a special direction: along the magnetization vector **M**. One can say the symmetry of the system changed from that of a sphere to that of a cylinder (Figure 10.3), and this change happens suddenly as soon as S is nonzero, no matter how small it is, so it does not cost any energy. These are all features of a second order phase transition.

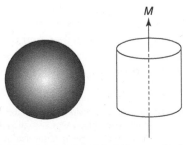

Figure 10.3 A sphere is more symmetrical than a cylinder

The state of the system with more symmetry is called 'disordered', and with less symmetry 'ordered phase'. This is not an error! In mathematics and physics symmetry means, essentially, that an object looks the same from different directions. The more such directions, the more symmetric is the object. A square is more symmetric than an equilateral triangle; a circle is more symmetric than a square; and a sphere is more symmetric than a cylinder.[3] In the same fashion, an empty piece of paper is more symmetric than a page with an intricate pattern.

From this point of view, a magnet above the Curie point is more symmetric: since there is no magnetization, all directions are the same. On the other hand, though there is more symmetry, there is less order. The magnetic moments randomly fluctuate, and at any given time you cannot tell in what direction a given vector s_j is more likely to point: its average is zero, so it can point in any direction with equal probability.

This is why the average magnetization (and similar quantities in other systems undergoing a second order phase transition) is called the *order parameter*. Above the transition temperature the order parameter is zero. Below the transition temperature it is finite.

Landau's theory of second order phase transitions and spontaneous symmetry breaking

So far we were talking about the *classical,* not quantum, model of a magnet. This is, strictly speaking, very inconsistent: as Bohr and van Leeuwen rigorously proved, magnetism only exists because of quantum mechanics. In classical physics there can be no 'atomic magnetic moments' μs_j and therefore no net magnetization M. (So, if you want to have a proof positive that quantum mechanics works ready at hand, carry a fridge magnet

---

3   The fact that a sphere (in three dimensions) and a circle (in two dimensions) are the most symmetric objects was recognized early on. This is why ancient philosophers envisaged a perfectly spherical Moon and perfectly spherical planets revolving around the Earth along circular trajectories.

in your pocket.) Nevertheless, if we assume that such quantum objects as atomic magnetic moments exist, we can for a while treat them as classical variables.

This is what Landau did in his theory of second order phase transitions. He developed it when working on the theory of liquid helium, and it proved to have a much wider applicability. Since then more accurate theories of phase transitions were developed, but it remains a very powerful, insightful and useful tool.

His idea was startlingly simple. It is known from classical thermodynamics and statistical mechanics that in equilibrium a physical system is in a state with the lowest *free energy*. Free energy is a certain thermodynamic function, which at absolute zero coincides with the total energy of the system.[4] Landau guessed that near the phase transition point (*critical temperature, T_c*) it would have the following form (Figure 10.4):

$$F(\phi) = a(T - T_c)\phi^2 + b\phi^4.$$

Here ϕ is the order parameter, which for simplicity we take to be a real non-negative number, same as a and b.

If $T > T_c$, both coefficients in the expression for $F(\phi)$ are positive, and the only minimum is at $\phi = 0$. In other words, above the critical temperature the order is absent.

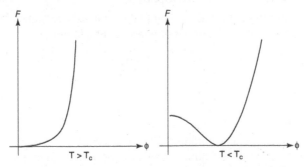

Figure 10.4 Free energy above and below the second order phase transition temperature

4   It has absolutely nothing to do with the 'free energy' to be extracted by various contraptions from vacuum fluctuations, zero-point energy, magnet power and what not, as promised by any number of crackpot websites.

The picture changes when $T < T_c$. As you can check, then the free energy has a maximum in the disordered state. The minimum is at $\phi = \sqrt{\frac{a}{2b}(T - T_c)}$ and this is the state in which the system will end up in equilibrium. Near the transition point the order parameter has characteristic (for the Landau theory) square root behaviour.

Now suppose that the order parameter can take both positive and negative values. This creates a problem: at $T < T_c$ there will be *two* minima of $F(\phi)$, at $\phi = \pm\sqrt{\frac{a}{2b}(T - T_c)}$. Which one to choose?

If we tried to calculate the order parameter by taking the average over the ensemble, we would obtain *zero*, because different systems in the ensemble would have positive or negative values of the order parameter at random. Both positive and negative terms would contribute equally and cancel each other. But in reality we would have just one system, with *either* positive *or* negative ϕ.

We have here encountered a very interesting and important phenomenon: *spontaneous symmetry breaking*.

Key idea: Equilibrium state

In an equilibrium state, according to the Landau theory of second-order phase transitions, the free energy is at a minimum.

The free energy has a minimum (or a maximum) if $\frac{dF}{d\phi} = 0$.

$$\frac{dF}{d\phi} = 2a(T - T_c)\phi + 4b\phi^3.$$

Check that
* if $T > T_c$, the only minimum $F(\phi)$ is at $\phi = 0$.
* If $T < T_c$, then $F(\phi)$ has a *maximum* at $\phi = 0$ and a minimum at

$$\phi = \sqrt{\frac{a}{2b}(T_c - T)}.$$

Spontaneous symmetry breaking occurs when the symmetry of a solution is lower than the symmetry of the problem.

What does this actually mean? In our case the problem – the function $F(\phi)$ – is symmetric: it does not depend on the sign of ϕ, only on its absolute value. The solution, on the other hand, is either $+\sqrt{\frac{a}{2b}(T-T_c)}$ or $-\sqrt{\frac{a}{2b}(T-T_c)}$, so that ϕ has a definite sign. The choice between them is made at random – any infinitesimally small perturbation at the transition point will push the system towards the one or the other.

A standard textbook illustration of spontaneous symmetry breaking is a sharp pencil balancing on its tip on a smooth horizontal surface: it can fall in any direction with equal probability, so the problem is symmetric, but the slightest whiff of air will bring it down, and it will end up pointing in one definite direction. A less standard one involves a mosquito and two heraldic animals carrying a heavy object.

Figure 10.5 Spontaneous symmetry breaking (Zagoskin 2014: 167, Fig. 4.5)

A similar situation arises in the Heisenberg magnet. There we can take for the order parameter the absolute value of the magnetization, $M = |\mathbf{M}|$, and write

$$F(M) = a(T - T_c)M^2 + bM^4.$$

When finding the average magnetization, we would run into the same problem: the ensemble-average $\langle M \rangle = 0$, because in different copies of our system within the ensemble the vector $\mathbf{M}$ can point in any direction. In order to go around it, Nikolay Bogoliubov suggested using the *'quasiaverages'*, $< M > \neq 0$.

They are found by first adding to the free energy a term
$-M \cdot h$ with a small h, calculating the average magnetization,
and then setting h exactly to zero. In this way we place the
system in the (imagined) external magnetic field h, and thereby
choose the direction in which it will prefer to magnetize in every
system within the ensemble. This is like telling the mosquito in
Figure 10.5 where to land.

From the classical Heisenberg magnet, it is a short trip to the
quantum Heisenberg magnet model. By replacing classical
vectors of fixed length with quantum angular momentum
operators we obtain the quantum Hamiltonian

$$\hat{H} = -\sum_{jk} J_{jk} \hat{s}_j \cdot \hat{s}_k - \mu \sum_j h \cdot \hat{s}_j,$$

which with minor modifications describes many natural
systems as well as quantum computers. For now we have
had enough of magnetic and magnet-like systems. Instead we
shall look at a system with a less obvious order parameter: a
superconductor.

Superconductors, superconducting phase transition and the superconducting order parameter

The time span between the discovery of superconductivity
in 1911 and its explanation in 1957 may seem excessive,
but only to a public accustomed to a steady and very fast
progress of physics over the last hundred years. When it
was discovered, quantum theory was still in its infancy, and
superconductivity is a macroscopic quantum phenomenon.
The theoretical description of superconductivity did not
appear all of a sudden in a finished form: there were a
number of intermediate steps, with stop-gap assumptions and
plausible guesses, and – remarkably – most of this guesswork
was eventually shown to follow from a consistent microscopic
theory.

Spotlight: Heike Kamerlingh Onnes

Superconductivity was discovered by Heike Kamerlingh Onnes (1853–1926) in the course of his experiments on the properties of matter near the absolute zero of temperature. These experiments were only made possible by his development and refinement of cryogenic technology. He received his Nobel Prize in Physics (1913) for this research, and in particular, for the liquefaction of helium. This was a well-deserved award: all cryogenics is based on the use of liquid helium (the common helium-4 and the rare isotope, helium-3), and costs of helium loom big in most physics departments' research budgets.

The first striking property of a system in a superconducting state was, of course, its ability to conduct electricity without resistance. This was what attracted the attention of Kamerlingh Onnes: his mercury sample, immersed in liquid helium, at 4.2 K started conducting electricity at zero voltage. This also gave the phenomenon its name.

It turned out that not all *normal* conductors become superconductors at low enough temperatures. Actually, the *best* normal conductors (like gold, silver or copper) *never* become superconducting and for a good reason, which we will discuss later. And usually the better a material is at superconducting, the worse it is at conducting electricity in normal state. For example, niobium becomes superconducting at 9 K, lead at 7 K, but aluminium – which is a good enough normal conductor to replace copper sometimes – becomes superconducting at slightly higher than 1 K.

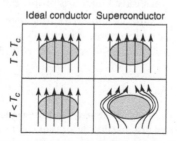

Figure 10.6 An ideal conductor vs. a superconductor: a superconductor expels the magnetic field

It also turned out that superconductors are not the same as 'ideal conductors' (conductors with zero electrical resistance). The difference is in their interaction with the magnetic field. Suppose that a piece of material, which becomes an ideal conductor below a certain temperature, is placed in the magnetic field at a higher temperature and then cooled down (Figure 10.6). The magnetic field will then be 'frozen' in the material. The reason is simple. Any change of the magnetic field will produce – by the virtue of Maxwell's equations (or, if you wish, Faraday's law of electrical induction) – electrical currents, which will flow in such a way as to produce a magnetic field of their own, which would completely cancel out any attempted change to the initial magnetic field. On the contrary, a superconductor does not freeze external magnetic fields: it expels them. This phenomenon, known as the *Meissner–Ochsenfeld effect*, discovered in 1933, played a significant role in the understanding of superconductivity. In particular, it helped bring about the realization that the transition to a superconducting state is a second-order phase transition.[5]

Key idea: The operator ∇

The operator ∇ (nabla), or *gradient*, indicates the magnitude and direction of the fastest growth of a function. For example,

$$\nabla f(r) = e_x \frac{\partial f}{\partial x} + e_y \frac{\partial f}{\partial y} + e_z \frac{\partial f}{\partial z}.$$

In 1950, Landau and Vitaly Ginzburg proposed the first consistent theory of superconductivity (the *Ginzburg–Landau theory*). It was the development of Landau's earlier theory of *superfluidity in liquid helium-4*, which we will touch upon later.

---

5 This applies if it takes place with zero external magnetic fields. If the system is initially placed in a magnetic field, as in Figure 10.6, this will be a first-order phase transition. The role of latent heat is played by the energy required to expel the magnetic field from the volume of space occupied by the superconductor.

The essence of this theory is contained in the expression for the free energy *density*[6] of a superconductor:

$$F(\psi(r)) - F_0 = a(T - T_c)|\psi(r)|^2 + b|\psi(r)|^4 + \frac{1}{2m^*}\left|\left(\frac{\hbar}{i}\nabla - \frac{q}{c}A\right)\psi(r)\right|^2 + \frac{B^2(r)}{8\pi}.$$

Here F_0 is the free energy density of the system in its normal state.

First, the right-hand side of this equation contains familiar terms, like in a magnet, only with the magnetic moment replaced by some *complex* function $\psi(r)$. This should be the order parameter in our system below the superconducting transition temperature T_c. The last term is (as you probably know from the theory of electromagnetism) the energy density of the magnetic field $B(r)$.[7] But what is the third term?

Key idea: Kinetic energy operators

The kinetic energy operator is $\frac{\hat{p}^2}{2m^*} = \frac{1}{2m^*}\left(\frac{\hbar}{i}\nabla\right)^2$. (This is another way of writing the expression for the momentum operator:

$\hat{p} = \frac{\hbar}{i}\left(e_x\frac{d}{dx} + e_y\frac{d}{dy} + e_z\frac{d}{dz}\right) = \frac{\hbar}{i}\nabla$). Then the expectation value of this operator in a state, described by a wave function $\psi(r)$ will be

$\int d^3r\,\psi^*(r)\frac{1}{2m^*}\left(\frac{\hbar}{i}\nabla\right)^2\psi(r) = \frac{1}{2m^*}\int d^3r\left|\frac{\hbar}{i}\nabla\psi(r)\right|^2$. Then, in Chapter 9 we saw that in the presence of magnetic field, described by a vector potential A, the quantum momentum operator $\hat{p}$ is replaced by $\hat{p} - \frac{q}{c}A$ for a particle with the electric charge q. Therefore the density of the kinetic energy in the system should indeed be

$\frac{1}{2m^*}\left|\left(\frac{\hbar}{i}\nabla - \frac{q}{c}A\right)\psi(r)\right|^2$.

Ginzburg–Landau's theory is a *phenomenological* theory. That is, it is content with a quantitative description of the observed

---

6   Free energy per unit volume. In order to find the free energy of the system, one must integrate this expression over all of space.
7   Again, we are using Gaussian units – your textbook is probably using the SI, so do not worry about the coefficients like $1/8\pi$ or $1/c$.

phenomena without trying to establish its relation to the first principles. Its authors assumed that the order parameter $\psi(r)$ is the 'macroscopic wave function of superconducting electrons', whatever this may mean. Then the third term makes sense: it is the density of the kinetic energy of such 'electrons'. Their charge and mass do not have to coincide with the charge and mass of free electrons in space, but rather serve as some effective parameters.[8] It turned out that $q = 2e$, twice the electron charge.

The 'superconducting wave function' is the order parameter of a superconductor. It is usually written as (up to a factor) as

$$\psi(r) \propto \Delta(r)e^{i\phi(r)},$$

separating its absolute value and its phase, Δ and ϕ. This makes sense. The absolute value determines the very existence of superconductivity: it is zero above T_c and nonzero below it. In the absence of external magnetic field, when both B and A are zero, we can assume that ψ is the same across the system, and therefore the 'kinetic' term in the free energy (which contains the derivative of ψ) will be zero as well. Then what remains is the same as in a magnet, with Δ playing the role of the absolute value of the magnetic moment, M, and nothing depends on the *superconducting phase* ϕ (as nothing depended on the direction of the magnetic moment).

Here we will again run into the same problem as with a magnet: since nothing depends on ϕ, it can take any value, and the ensemble average of the complex order parameter ψ would be zero. Again, this is resolved by using quasiaverages.

The phase of the order parameter of an isolated superconductor has no physical meaning – much as the overall phase of a wave function. But its gradient – the magnitude and direction of its change in space – plays an important role. It determines the magnitude and direction of the *supercurrent* – the very current that flows without resistance and that attracted attention of Kamerlingh Onnes in the first place.

---

8   It follows from other considerations that the 'charge' q should be an *integer multiple* of the elementary charge e.

Key idea: Velocity operators

The velocity operator is $\dfrac{\hat{p}}{m^*} = \dfrac{1}{m^*}\dfrac{\hbar}{i}\nabla$. The expectation value of velocity at the point $\boldsymbol{r}$ is then $\psi^*(\boldsymbol{r})\dfrac{1}{m^*}\dfrac{\hbar}{i}\nabla\psi(\boldsymbol{r})$. Assuming that $\psi(\boldsymbol{r}) \propto \Delta e^{i\phi}$ and Δ is a constant, we see that the velocity is $v \propto \dfrac{\hbar}{m^*}\nabla\phi$, and the electric current density $j \propto \dfrac{q\hbar}{m^*}\nabla\phi$. For example, $j_x \propto \dfrac{q\hbar}{m^*}\dfrac{d\phi}{dx}$.

The Ginzburg–Landau theory is a very good theory. It nicely describes the behaviour of many superconductors near the superconducting transition temperature. It reproduces from a single point of view the earlier, ad hoc, equations, which correctly described certain properties of superconductors. In particular, it enables the calculation of supercurrents in real systems and explains the behaviour of superconductors in the magnetic field, like the Meissner–Ochsenfeld effect.[9] But, being a phenomenological theory, it does not allow us to predict the properties of a given material (will it become superconducting? at what temperature? and so on), and it does not explain which order parameter it manipulates with. A microscopic theory was still badly needed.

---

9   This is an interesting explanation: magnetic (and in general, electromagnetic) fields cannot penetrate the superconductor, because photons, due to their interaction with the order parameter $\psi(r)$, 'acquire mass' (that is, behave as if they were massive particles). The fact that photons in a superconductor become massive quasiparticles is in itself no more surprising than electrons in a metal having an effective mass different from that in vacuum, or that the Coulomb potential in a metal is screened. But it, along with a number of other effects in condensed matter theory, provided useful analogies and directions to, for example, particle physics and cosmology.

BCS theory, Bose–Einstein condensate, Cooper pairs and the superconducting energy gap

The *BCS (Bardeen-Cooper-Schrieffer) theory* of superconducting state was based on a seemingly absurd assumption: that in superconductors the conduction electrons weakly attract each other. Of course, those electrons are quasiparticles, and we already know that the electrostatic repulsion between them is drastically modified (screened), but *attraction*? We will get to justify this assumption in a moment, but first let us see what this assumption leads to, and why it was made in the first place.

Spotlight: Bardeen, Cooper and Schrieffer

John Bardeen, Leon Cooper and John Robert Schrieffer developed their BCS theory of superconductivity in 1957 and received the Nobel Prize in Physics for it in 1972.

Bardeen's achievement is all the more remarkable, since just a year before his discovery (in 1956) he shared with William Shockley and Walter Brattain another Nobel Prize in Physics for the invention of the transistor. He therefore played a leading role in two of the three main directions of the 'First Quantum Revolution' in technology (semiconductors, superconductors and lasers).

Independently, the essentially identical theory of superconductivity was proposed by Nikolay Nikolayevich Bogolyubov.

The reason for making the assumption is quantum statistics – the fundamental difference between bosons and fermions. Fermions cannot occupy the same quantum state, while bosons can, in any number. The ultimate manifestation of this difference is the *Bose–Einstein condensation*.

For bosons the average number of particles in a state with energy E is given by the Bose formula,[10]

---

10   For the fermions, as you recall, there was +1 instead of –1 in the denominator. What a difference a sign can make!

$$n_B(E; \mu, T) = \frac{1}{\exp\left(\frac{E-\mu}{K_B T}\right) - 1}$$

Key idea: Photons

What about photons? The Planck formula is the Bose formula for $\mu = 0$! Fortunately, there is no problem here. Photons can be emitted or absorbed at will, their number is not conserved, and the requirement $\mu < 0$ does not apply.

Here μ is the *chemical potential*, the counterpart of the Fermi energy for fermions, and it is determined by the condition that the total number of particles in the system is fixed:

$$\sum_E n_B(E; \mu, T) = N.$$

Clearly, for any E the number $n_B(E)$ must be positive: a negative number of particles is not something even quantum mechanics can live with. Therefore $e^{\frac{E-\mu}{K_B T}} - 1 > 0$, that is, $e^{\frac{E-\mu}{K_B T}} > 1$ for *any* energy $E \geq 0$. Therefore μ is always negative (compare this to the always positive Fermi energy).

In a large system, where the energy levels are very dense, the sum $\sum_E n_B(E; \mu, T)$ can be usually rewritten as an integral, $\int_0^\infty \rho(E) n_B(E; \mu, T)\, dE$. Here $\rho(E)$ is the density of states – in a small energy interval ΔE there are $\rho(E)\Delta E$ energy levels. Here 'usually' is not a figure of speech: the condition

$$\int_0^\infty \rho(E) n_B(E; \mu, T)\, dE = N$$

sometimes cannot be satisfied with any $\mu < 0$ (or even $\mu = 0$). This is a mathematical fact: if the density of states drops to zero as $E \to 0$ (which is the case in three-dimensional systems), then below certain temperature T_c the normalization condition cannot be met. What happens then is the Bose–Einstein condensation:

a macroscopic number of particles will congregate in the
ground state of the system, with zero energy. The normalization
condition then becomes

$$N_0(T) + \int_0^\infty \rho(E) n_B(E;0,T)\,dE = N.$$

Key idea: Bose–Einstein condensation

The Bose–Einstein condensation temperature T_c is determined
from the condition

$$\int_0^\infty \rho(E) n_B(E;0,T)\,dE = \int_0^\infty \frac{\rho(E)}{e^{\frac{E}{k_B T_c}} - 1}\,dE = N$$

You can't fall off the floor.
Paul's Law

All the N_0 particles, which cannot be 'accommodated' in the
states with positive energies, are falling to the bottom, to the
ground state, forming the *Bose–Einstein condensate*. The lower
the temperature, the more of them will be in the condensate.
(Do not misunderstand me: all quantum particles of the same
kind are indistinguishable. You cannot point your finger at a
particular boson and say: *this* one is in the condensate. We are
talking here about the occupation numbers of states with given
energies, but to say just 'particle' saves time and space – as long
as you remember what is actually meant here.)

Spotlight: Kapitsa, Allen and Misener

The superfluidity of helium-4 was discovered in 1937 by Pyotr
Leonidovich Kapitsa (1894–1984) (Nobel Prize in Physics, 1978) and
independently by John F Allen and Don Misener.

Kapitsa worked in Rutherford's Cavendish Laboratory in
Cambridge for a decade, where he achieved important discoveries

Now, what all this has to do with superconductors? The key feature of the condensate is that it is in the ground state. It means that it cannot possibly decrease its energy any more. If the condensate can be brought to motion as a whole, then its energy is, of course, higher than that of a resting condensate, but certain rather strict conditions must be met for the moving condensate to dissipate its energy and stop. Therefore such a motion can continue indefinitely. The strange properties of *superfluid helium-4* below 2.172 K are explained by the presence of the Bose condensate of helium atoms. (These properties are really strange: for example, superfluid helium-4 freely flows through narrowest capillaries; if put in an open vessel, a thin film of liquid helium will creep up the wall and down the other side and then out, and eventually all of the superfluid helium will gather in a pool at the lowest place of the apparatus; and so on and so forth.) It was reasonable to speculate that a condensate of electrically charged particles – for example, electrons – could carry electric supercurrent. But electrons are *fermions*!

On the other hand, though the atoms of helium-4 are bosons, they consist of fermions: their nuclei (α-particles) contain two protons (fermions!) and two neutrons (fermions!), and the two electrons of helium are also fermions. What matters is the total spin of the particle – as we know, if it is an integer multiple of $\hbar$, the particle is a boson (in the case of a helium-4 atom, a *compound* boson).

Therefore if two electrons could form a pair with total spin equal to $n\hbar$, these pairs should be able to form a *superconducting* Bose–Einstein condensate. This was the guiding idea in the search for the microscopic theory of superconductivity.

The key element of the BCS theory is the formation of *Cooper pairs* of electrons with *opposite* spins. (Then the total spin is zero, and Cooper pairs are *scalar* compound bosons.)

The BCS theory (and Nature) has some interesting subtleties. In principle, in the case of strong enough electron–electron attraction, electrons could form 'two-electron molecules' in real space already above the superconducting transition temperature. Then, as temperature is lowered, these 'molecules' could undergo a superconducting phase transition and form a condensate, the same way as helium-4 atoms do during the superfluid phase transition. Such theories were actually developed. But this happened not to be the case – at least in usual superconductors.

The Cooper pairs form in the *momentum space*, and this happens if there is an arbitrarily weak attraction between electrons in a narrow layer near the Fermi surface. Electrons with momentum p and spin 'up' and with momentum $-p$ and spin 'down' become correlated – they form a bound state and this happens exactly at the point of superconducting phase transition. In BCS theory the bosons and the Bose–Einstein condensate appear *simultaneously*.

Mathematically it is manifested in the appearance of nonzero *anomalous averages*: below the transition temperature T_c the ground state average

$$\left\langle c_{p\uparrow}c_{-p\downarrow}\right\rangle_0 = \left\langle \Psi_0 \left| c_{p\uparrow}c_{-p\downarrow} \right| \Psi_0\right\rangle \neq 0.$$

If you have no objections against this equation, then you clearly did not pay attention when reading Chapter 7. The state $c_{p\uparrow}c_{-p\downarrow}|\Psi_0\rangle$ obviously contains two less electrons than the state $|\Psi_0\rangle$, and therefore they must be orthogonal: $\left\langle \Psi_0 \right| \cdot \left(c_{p\uparrow}c_{-p\downarrow}|\Psi_0\rangle \right) = \left\langle \Psi_0 \left| c_{p\uparrow}c_{-p\downarrow} \right| \Psi_0 \right\rangle = 0!$ Nevertheless the equation is right – here lies one of the subtleties of the BCS theory (and Nature). The superconducting condensate, described by $|\Psi_0\rangle$, contains so huge a number of electrons, that two more or less does not really matter. The appearance of nonzero anomalous averages in the ground state of a system is the signature of the condensate.

Key idea: Macroscopic wave function

The anomalous average in real space is the very 'macroscopic wave function of superconducting electrons' $\psi(r)$, which was introduced by Landau and Ginzburg in their theory:

$$\psi(r) \propto \left\langle \hat{\psi}_\uparrow(r)\hat{\psi}_\downarrow(r) \right\rangle_0.$$

We have here switched from the annihilation operators c_p in the momentum space to the field operators $\hat{\psi}(r)$ in real space.

The subsequent analysis of the BCS theory showed that it reproduces the Ginzburg–Landau theory close to the transition temperature. But it also worked far from it; it explained the actual nature of superconductivity; it identified the condensate of Cooper pairs as its key component; and it provided a *qualitative* connection between the strength of the electron–electron attraction and the superconducting transition temperature.

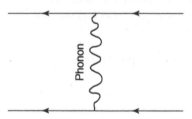

Figure 10.7 Electron–electron interaction via a phonon (a Feynman diagram)

Still, where did the attraction come from? A Feynman diagram gives us the answer (Figure 10.7). The crystal lattice, in which the electrons move, is not infinitely rigid: ions oscillate around their equilibrium positions. The quanta of these oscillations are, as we have mentioned, called *phonons* – they are essentially quantized sound waves. An electron, moving through the lattice, leaves behind a wake of ion disturbances: they are attracted by the electron, but are too heavy to react at once. Therefore quite long after the electron is gone, there remains a region with an extra concentration of ions and, therefore, net positive electric charge. This region can attract another electron and so an effective *electron–electron attraction mediated by phonons* is produced!

This effective electron–electron coupling can be taken into account by adding to the Hamiltonian the term $\int d^3 r \left[-g\hat{\psi}_\uparrow^+ \hat{\psi}_\downarrow^+ \hat{\psi}_\downarrow \hat{\psi}_\uparrow \right]$ (g is a positive constant), and further approximated by writing

$$\int d^3 r \left[-g\hat{\psi}_\uparrow^+ \hat{\psi}_\downarrow^+ \hat{\psi}_\downarrow \hat{\psi}_\uparrow \right] \approx \int d^3 r \left[-g\langle \hat{\psi}_\uparrow^+ \hat{\psi}_\downarrow^+ \rangle_0 \hat{\psi}_\downarrow \hat{\psi}_\uparrow \right] + \int d^3 r \left[-g\,\hat{\psi}_\uparrow^+ \hat{\psi}_\downarrow^+ \langle \hat{\psi}_\downarrow \hat{\psi}_\uparrow \rangle_0 \right] =$$
$$\int d^3 r \left[-\Delta^*(r)\hat{\psi}_\downarrow \hat{\psi}_\uparrow - \Delta(r)\hat{\psi}_\uparrow^+ \hat{\psi}_\downarrow^+ \right].$$

The term $\Delta(r) = g\langle \hat{\psi}_\downarrow \hat{\psi}_\uparrow \rangle_0$ is called the *pairing potential* and – up to the factor g – coincides with the superconducting order parameter (if include in it the phase factor $\exp(i\varphi)$. It is a complex-valued function of position in real space, r. Its special importance is that its absolute value determines the magnitude of the *superconducting energy gap* – that is, the minimal energy that can be taken from the superconducting condensate. This can be only done by breaking up a Cooper pair and one must provide *each* of its electrons at least energy $|\Delta(r)|$ in order to do that. Therefore indeed one cannot easily stop a moving condensate – and supercurrents can flow undisturbed.

Spotlight: The God particle

The appearance of the Bose–Einstein condensate, which drastically changes the properties of the system, is a powerful idea and was used in many other areas of physics. Take, for example, the celebrated Higgs boson (called 'the God particle' not because of some piety on the physicists' side, but because for many years 'it was like God: discussed by everybody and seen by no one'). The nonzero vacuum value of the Higgs field (the Higgs condensate) provides masses to the three bosons responsible for the weak interaction in a manner, similar to how the superconducting order parameter makes photons in a superconductor 'massive' and leads to the Meissner–Ochsenfeld effect.

An important note: since electromagnetic fields cannot penetrate inside a superconductor, and electric currents produce magnetic fields, the supercurrent must flow in a thin layer near the surface of a superconductor. (The thickness of this layer is called the *superconducting penetration depth*). In the bulk of a superconductor there are no currents and no fields.

Josephson effect*

Now let us consider a beautiful and instructive phenomenon, which has many applications in science and technology. This is the *Josephson effect*.

We have seen that the Ginzburg–Landau theory was put on a solid microscopic foundation. Its superconducting order parameter Ψ (*r*) can indeed be considered as a macroscopic wave function of the superconducting condensate of Cooper pairs. The effective mass m^* is twice the mass of an electron (a *conduction* electron, not the one in vacuum), and the charge $q = 2e$ is twice the electron charge. What is more to the point, it is the condensate density, that is, the anomalous average, $\langle \hat{\psi}_\downarrow \hat{\psi}_\uparrow \rangle_0$, and not the energy gap, $|g \langle \hat{\psi} \downarrow \hat{\psi} \uparrow \rangle_0|$, that determines the superconductivity. The coupling strength g, and with it the energy gap, can be zero in some part of the system, but there will still be superconductivity there.

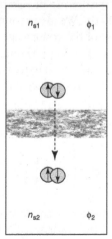

Figure 10.8 Josephson effect

This was what Josephson realized. He considered two bulk pieces of superconductor separated by a thin insulating barrier, through which electrons could tunnel: the so-called *Josephson junction* (Figure 10.8). The common wisdom then was that superconducting current could not flow through, since it required the tunnelling of two electrons at once. If the probability of one electron tunnelling is $P \ll 1$ the probability of two simultaneous tunnelling should be P^2, which is vanishingly small. But if superconducting condensate can exist in the barrier (where there is no electron–electron attraction and therefore $g = 0$ and the gap is zero), then the Cooper pair can tunnel as a single quantum particle with a small, but not negligible probability P.

Here we will follow a simple approach proposed by Feynman. He suggested describing the Josephson junction by a two-component wave function,

$$\Psi = \begin{pmatrix} \sqrt{n_{s1}}\, e^{i\phi_1} \\ \sqrt{n_{s2}}\, e^{i\phi_2} \end{pmatrix}.$$

The components are Ginzburg–Landau condensate wave functions, which are normalized in such a way that their modulus squared gives the density of electrons in the condensate. This is the usual thing to do. We also assume that neither the magnitude nor the phase of the condensate change inside bulk superconductors. This is a good approximation: the possible flow of Cooper pairs through the barrier is a mere trickle (since $P \ll 1$), and therefore it will not noticeably disturb the bulk.

Now, the Hamiltonian of the junction, acting on Ψ, can be written as a two-by-two matrix,

$$\hat{H} = \begin{pmatrix} 0 & K \\ K & 0 \end{pmatrix},$$

where K describes the transfer of an electron across the barrier. It is proportional to the barrier transparency P.

The Schrödinger equation will be then

$$i\hbar \frac{\partial}{\partial t}\, \Psi = i\hbar \begin{pmatrix} \left\{\dfrac{\dot{n}_{s1}}{2\sqrt{n_{s1}}} + i\dot{\phi}_1\right\} e^{i\phi_1} \\[2ex] \left\{\dfrac{\dot{n}_{s2}}{2\sqrt{n_{s2}}} + i\dot{\phi}_2\right\} e^{i\phi_2} \end{pmatrix} = K \begin{pmatrix} \sqrt{n_{s2}}\, e^{i\phi_2} \\[2ex] \sqrt{n_{s1}}\, e^{i\phi_1} \end{pmatrix}.$$

If the phases of superconductors are time-independent, then we obtain two equations:

$$\frac{\partial n_{s1}}{\partial t} = \frac{2K}{i\hbar}\sqrt{n_{s1}n_{s2}}\, e^{i(\phi_2 - \phi_1)}; \frac{\partial n_{s2}}{\partial t} = \frac{2K}{i\hbar}\sqrt{n_{s1}n_{s2}}\, e^{i(\phi_1 - \phi_2)}.$$

Let us subtract the one from the other and recall that $\sin x = (e^{ix} - e^{-ix})/2i$. We will find, that

$$\frac{\partial}{\partial t}(n_{s2} - n_{s1}) = \frac{4K}{\hbar}\sqrt{n_{s1}n_{s2}}\sin(\phi_1 - \phi_2).$$

What does it mean? The change in the number of electrons in one superconductor is due to their flow through the barrier to the other superconductor. Therefore, if we multiply the left-hand side by the electron charge, we will obtain (twice) the electric current through the barrier. We can therefore rewrite the last equation as

$$I(\phi_1 - \phi_2) = I_c \sin(\phi_1 - \phi_2).$$

This is the formula for the *DC Josephson effect*.[11] It tells us that between two superconductors separated by a thin enough insulating barrier (or, generally, by any *weak link*, which impedes the transfer of electrons between the two: a narrow contact, a thin layer of non-superconducting metal, etc.) a superconducting current can flow *in equilibrium*, as long as there is a superconducting phase difference between them. There is no resistance. There is no voltage. The maximal amplitude of

---

11   There is also the *AC Josephson effect*: if there is a voltage difference between the superconductors, the phase difference between them will oscillate with a frequency, proportional to the voltage. You may try to derive it from these equations.

this current, I_c, is called the *Josephson critical current*. Note that it is proportional to K (i.e. to the transparency of the barrier to a single electron) and not to K^2, as everybody expected before Josephson.

This is a beautiful science. The Josephson effect was quickly confirmed in experiment and its analogues in other areas have been studied ever since. In technology, the Josephson junctions are key components of SQUIDs (devices used as extremely sensitive magnetometers – e.g. with applications in the medical imaging, which we have discussed earlier). They are used for voltage standards, etc., but one of the most interesting and challenging applications of Josephson junctions is towards quantum computation. The superconducting quantum bits of all kinds (there are several of them) contain Josephson junctions and rely on the Josephson effect for their operation.

Fact-check

1 The adjacent atomic magnetic moments tend to point in the same direction, if the coupling is

   a antiferromagnetic
   b ferromagnetic
   c diamagnetic
   d paramagnetic

2 Point out an example of a second order phase transition

   a freezing
   b magnetization
   c melting
   d evaporation

3 What system properties change during the second order phase transition?

   a density
   b symmetry
   c volume
   d electric charge

4 Spontaneous symmetry breaking means that

   a   the symmetry of the system is randomly destroyed

   b   the system undergoes a phase transition

   c   the symmetry of the system's state is lower than the symmetry of the equations which govern its behaviour

   d   the system becomes more symmetric

5 One can distinguish a superconductor from an ideal conductor, because

   a   an ideal conductor has zero electrical resistance

   b   an ideal conductor expels the external magnetic field

   c   a superconductor expels the external magnetic field

   d   a superconductor freezes in an external magnetic field

6 Free energy

   a   is the energy that can be freely extracted from a physical system

   b   is a quantum mechanical operator of a physical system

   c   coincides with the energy of a system at absolute zero

   d   is a thermodynamic function, which is at a minimum when the system is in equilibrium

7 The order parameter of a superconductor is

   a   a Hermitian operator

   b   a complex number

   c   an expectation value of two field operators

   d   zero above the transition temperature

8 During the Bose–Einstein condensation

   a   all bosons occupy the lowest energy state

   b   many bosons occupy the lowest energy state

   c   the Bose–Einstein condensate appears

   d   the occupation number of the ground state becomes macroscopic

9 Superconductivity is possible, because

    a   electrons in a conductor can undergo the Bose–Einstein condensation

    b   electrons in a conductor can form Cooper pairs

    c   electrons in a conductor interact with phonons

    d   electrons are fermions

10 The Josephson effect is possible, because

    a   a Cooper pair tunnels coherently

    b   tunnelling amplitude for two electrons is two times less than that for one electron

    c   there is no voltage across the tunnelling barrier

    d   tunnelling probability for two electrons is two times less than that for one electron

Dig deeper

F. Reif, *Statistical Physics* (Berkeley Physics Course, Vol. 5). McGraw-Hill Book Company, 1967.

For an introduction to superconductivity (its science and scientists) see

J. Matricon and G. Waysand, *The Cold Wars: A History of Superconductivity*. Rutgers University Press, 2003.

V. V. Schmidt, *The Physics of Superconductors: Introduction to Fundamentals and Applications*. Springer, 1997, 2010.

11

Curiouser and curiouser

One of the main objections against Newton's theory of gravity was that it introduced *action at a distance*: in order for the law of inverse squares to work, any two gravitating masses had to 'know' the positions of each other at any instant, no matter how large a distance separated them. The adherents of Descartes' physics, which dominated European natural philosophy before Newton and did not go down without a fight, pointed that out as the cardinal flaw of the new fangled theory. To them action at a distance was inconceivable, and gravity was supposed to be the result of interaction of bodies with the aether – an all-pervading medium. The vortices in aether were supposed to be responsible for the formation of planetary orbits and for the fall of heavy bodies.

Unfortunately, Cartesian physics itself had a cardinal flaw: paraphrasing Laplace, it could explain anything and predict nothing. Newtonian mechanics, despite (or rather thanks to) its 'unphysical' action at a distance could and did make quantitative predictions, and that sealed the matter for the time. The Coulomb law was another version of action at a distance. Only when the attempts to present electrodynamics in the same way ran into serious problems in the mid-19th century while Maxwell's theory was successful, the action at a distance theories started going out of fashion. When the general relativity took over from the Newtonian gravity, it seemed that *local* theories would rule forever. And then came quantum mechanics...

Einstein-Podolsky-Rosen paradox

The intrinsic randomness of quantum theory was, of course, an irritant for the physicists accustomed to the (theoretically strict) Laplacian determinism of classical mechanics. But this irritation was mitigated by the experience of classical statistical mechanics and the *practical* impossibility to follow on this determinism for a macroscopic system (for example, a roomful, or even a fraction of the cubic millimetre, of gas). It seemed that quantum mechanics could well be an approximation, which can be derived from a 'more complete' theory. Such a theory would be deterministic and operate with some quantities ('hidden variables'), which for some reason only manifest themselves through quantum mechanical observables. There were precedents. For example, equations of the theory of elasticity were known and accurately described the properties of solids long before the atomic structure of matter was established, not to speak about the observation of atoms themselves.

Spotlight: Boltzmann and Loschmidt

In the famous exchange between Ludwig Boltzmann and Johann Loschmidt, the latter argued that if one would at a certain moment reverse the velocities of all the molecules in a volume of gas, then, following the laws of classical mechanics, they would exactly retrace their trajectories backwards and the gas would return to its initial state. For example, it would go back into a canister and compress itself, thus contradicting Boltzmann's *H-theorem* of statistical mechanics (according to which this could never happen). Boltzmann's response to this argument (known as the *Loschmidt's paradox*) was: 'Go on, reverse them!'

These hopes were dashed when it was firmly established that not only quantum mechanics (that is, Nature) is random, but it is *non-local* as well.

The non-locality could be inferred from, e.g. the double-slit experiments with single particles: the pattern on the screen depends on the particle being able to go through *both* slits. It is clear in Feynman's path integrals formulation of quantum mechanics. It is obvious when you look at the Aharonov–Bohm effect, when the behaviour of a particle depends on the magnetic field, though the particle cannot even in theory travel through the region where this field is present. But long before Feynman's and Aharonov–Bohm's work, in 1935, Einstein, Podolsky and Rosen invented a Gedankenexperiment (the *EPR experiment*), which, as they thought, established that quantum mechanics is incomplete – that is, must eventually reduce to some theory of hidden variables. Anything else would be inconceivable.

Figure 11.1 The Einstein-Podolsky-Rosen experiment

The idea of the experiment is quite simple (Figure 11.1), though it was not presented in this particular way in the EPR's original paper. Suppose that at a point O there is a pair of spin-½ particles in a singlet spin state (that is, with opposite spins and therefore zero total spin – like a Cooper pair), and at points A and B there is measuring apparatus, i.e. detectors, which can measure the component of a particle spin along certain axis. Let the particles fly in the opposite directions with high speeds v_A, v_B, such that $v_A + v_B > c$ (speed of light in vacuum), and impinge on the detectors. The spins in such a two-particle system are completely separated from other observables (such as positions or momenta), and its spin quantum state can be written as, e.g. $|EPR\rangle = \dfrac{|\uparrow\downarrow\rangle - |\downarrow\uparrow\rangle}{\sqrt{2}}$,

where $\downarrow(\downarrow)$ denotes the spin projection +1/2 (−1/2) in the direction Oz. Of course, we could choose any other axis: Ox, Oy or any direction whatsoever.

A boring, but instructive, exercise

Let us check that for the system in state $|\psi\rangle$ the expectation value of the projection of its total spin on any axis (Ox, Oy, or Oz) is indeed zero. The operator of a total spin projection on the Ox for a system of two spin-1/2 particles is $\hat{S}_x = \frac{1}{2}\left(\sigma_x^A \otimes \sigma_0^B + \sigma_0^A \otimes \sigma_x^B\right)$, and similar for the other two axes. (We measure spin in the units of $\hbar$). Here $\sigma_{x,y,z}$ is one of 2-by-2 Pauli matrices, and σ_0 is a unit 2-by-2 matrix (as explained in Chapter 4). The symbol $\otimes$ (called the *direct product*) stresses that a given operator (matrix) acts only on the quantum state of 'its own' particle (indicated by the superscript).

We can choose the coordinates in such a way that the 'up' state (along the axis Oz) is the state $\begin{pmatrix} 1 \\ 0 \end{pmatrix}$, and the 'down' state is $\begin{pmatrix} 0 \\ 1 \end{pmatrix}$. Then the expectation value of $\hat{S}_x$ in the state $\left|EPR\right\rangle = \frac{|\uparrow\downarrow\rangle - |\downarrow\uparrow\rangle}{\sqrt{2}}$ is (figure out where the $\frac{1}{4}$ came from!)

$$\left\langle EPR\left|\hat{S}_x\right|EPR\right\rangle = \frac{1}{4}\left(\left\langle\uparrow\downarrow - \downarrow\uparrow\left|\sigma_x^A \otimes \sigma_0^B + \sigma_0^A \otimes \sigma_x^B\right|\uparrow\downarrow - \downarrow\uparrow\right\rangle\right).$$

Take the $\sigma_x^A \otimes \sigma_0^B$ -term first.

$$\left\langle\psi\left|\sigma_x^A \otimes \sigma_0^B\right|\psi\right\rangle = \frac{1}{4}\left\langle\uparrow\downarrow - \downarrow\uparrow\left|\sigma_x^A \otimes \sigma_0^B\right|\uparrow\downarrow - \downarrow\uparrow\right\rangle = \frac{1}{4}\left\langle\uparrow\downarrow\left|\sigma_x^A \otimes \sigma_0^B\right|\uparrow\downarrow\right\rangle - \frac{1}{4}\left\langle\uparrow\downarrow\left|\sigma_x^A \otimes \sigma_0^B\right|\downarrow\uparrow\right\rangle$$
$$- \frac{1}{4}\left\langle\downarrow\uparrow\left|\sigma_x^A \otimes \sigma_0^B\right|\uparrow\downarrow\right\rangle + \frac{1}{4}\left\langle\downarrow\uparrow\left|\sigma_x^A \otimes \sigma_0^B\right|\downarrow\uparrow\right\rangle = \frac{1}{4}\left\langle\uparrow\left|\sigma_x^A\right|\uparrow\right\rangle\left\langle\downarrow\left|\sigma_0^B\right|\downarrow\right\rangle$$
$$- \frac{1}{4}\left\langle\uparrow\left|\sigma_x^A\right|\downarrow\right\rangle\left\langle\downarrow\left|\sigma_0^B\right|\uparrow\right\rangle - \frac{1}{4}\left\langle\downarrow\left|\sigma_x^A\right|\uparrow\right\rangle\left\langle\uparrow\left|\sigma_0^B\right|\downarrow\right\rangle + \frac{1}{4}\left\langle\downarrow\left|\sigma_x^A\right|\downarrow\right\rangle\left\langle\uparrow\left|\sigma_0^B\right|\uparrow\right\rangle.$$

We have here explicitly written that σ_x^A only acts on the first 'arrow' (particle A), and σ_0^B only on the second (particle B). The matrix σ_0^B is a unit matrix. It does not change the state it acts upon: for example,

$$\sigma_0\left|\uparrow\right\rangle = \begin{pmatrix} 1 & 0 \\ 0 & 1 \end{pmatrix}\begin{pmatrix} 1 \\ 0 \end{pmatrix} = \begin{pmatrix} 1 \\ 0 \end{pmatrix} = \left|\uparrow\right\rangle.$$

Since $\langle\uparrow|\downarrow\rangle = 0$ and $\langle\downarrow|\uparrow\rangle = 0$, we can immediately drop all the terms with $\langle\uparrow|\sigma_0^B|\downarrow\rangle = \langle\uparrow|\downarrow\rangle = 0$ and $\langle\downarrow|\sigma_0^B|\uparrow\rangle = \langle\downarrow|\uparrow\rangle = 0$.

Let us align both detectors with the axis Oz. Suppose the apparatus B is a little farther from O, or that the particle B flies a little slower, so that the detector A always fires first.

According to the rules of quantum mechanics, during measurement the quantum state of the system is randomly projected on one of the eigenstates of the operator being measured, and the measured value will be the corresponding eigenstate. The operator measured by the detector A is (we measure spins in the convenient units of $\hbar$) $\hat{S}_z^A=\frac{1}{2}\sigma_z^A\otimes\sigma_0^B$. The $\otimes\sigma_0^B$ -term simply means that the detector A only measures the spin of particle A and does not do anything to particle B. (It cannot: the interaction of the particle with a detector is local, and the detectors are so far apart that even light cannot reach B from A before the particle A is measured.) The eigenstates of σ_z are $|\uparrow\rangle=\begin{pmatrix}1\\0\end{pmatrix}$ with the eigenvalue +1 and $|\downarrow\rangle=\begin{pmatrix}0\\1\end{pmatrix}$ with the eigenvalue -1. The initial state of the system was $|EPR\rangle=\frac{|\uparrow\downarrow\rangle-|\downarrow\uparrow\rangle}{\sqrt{2}}$. Therefore after measurement it will be either $|\uparrow\downarrow\rangle$ (if the measured spin projection S_z^A is +1/2) or $|\downarrow\uparrow\rangle$ (if S_z^A is -1/2). Either outcome has the probability ½.

Now an important point: $\hat{s}^A+\hat{s}^B=0$, the total spin of the system is zero. Therefore *at the same moment*, when the apparatus A measures the spin s_z^A of particle A, the spin of particle B becomes fully determined: $s_z^B=-s_z^A$. If $s_z^A=\frac{1}{2}$, then $s_z^B=\frac{1}{2}$, and vice versa.

Figure 11.2 No pirates were hurt during this experiment

So far so good: we have randomness (the probabilities of observing $s_z^A = \frac{1}{2}$ or $s_z^A = -\frac{1}{2}$ are equal), but this could be a garden-variety classical randomness. Instead of sending two particles we could send two pirates, giving one of them a black mark and the other a white mark in sealed envelopes, chosen at random, with strict orders to open the envelopes only when reaching the points A and B (Figure 11.2). When the pirate A opens his envelope, he instantaneously realizes what mark his comrade has, but this only concerns the state of his knowledge. The situation was fixed irreversibly the moment when the two envelopes were randomly distributed at the starting point, and not when the pirates reached their destinations.

But this is not the case in quantum mechanics. Let us see what happens if, while the particle A is still in flight, we turn the detector 90°, to let it measure the spin projection on the axis Ox.

The eigenstates of the operator $\hat{s}_x^A = \frac{1}{2}\sigma_x^A = \frac{1}{2}\begin{pmatrix} 0 & 1 \\ 1 & 0 \end{pmatrix}$ are $|\rightarrow\rangle = \frac{1}{\sqrt{2}}\begin{pmatrix} 1 \\ 1 \end{pmatrix}$ with the eigenvalue ½ and $|\leftarrow\rangle = \frac{1}{\sqrt{2}}\begin{pmatrix} 1 \\ -1 \end{pmatrix}$ with the eigenvalue –½.

The states $|\uparrow\rangle$ and $|\downarrow\rangle$ can be written as

$$|\uparrow\rangle = \frac{|\rightarrow\rangle + |\leftarrow\rangle}{\sqrt{2}}; |\downarrow\rangle = \frac{|\rightarrow\rangle - |\leftarrow\rangle}{\sqrt{2}}.$$

This means that our EPR-state can be rewritten as

$$|EPR\rangle = \frac{1}{\sqrt{2}}\left\{\frac{|{\rightarrow}{\downarrow}\rangle + |{\leftarrow}{\downarrow}\rangle}{\sqrt{2}} - \frac{|{\rightarrow}{\uparrow}\rangle - |{\leftarrow}{\uparrow}\rangle}{\sqrt{2}}\right\} = \frac{1}{\sqrt{2}}\left\{\frac{|{\rightarrow}{\downarrow}\rangle - |{\rightarrow}{\uparrow}\rangle}{\sqrt{2}} + \frac{|{\leftarrow}{\downarrow}\rangle + |{\leftarrow}{\uparrow}\rangle}{\sqrt{2}}\right\}.$$

If the measurement of the x-component of spin of the particle A yields $+1/2$, then this particle after the measurement is in the state $|{\rightarrow}\rangle$. Investigating the above formula for $|EPR$, we see that the probability of this outcome is ½, and that the quantum state of the particle B after this measurement of the particle A will be

$$\frac{|{\downarrow}\rangle - |{\uparrow}\rangle}{\sqrt{2}} = -|{\leftarrow}\rangle,$$

that is, the eigenstate of the operator $\hat{s}_x^B$ with the eigenvalue $-1/2$.[1] Therefore, if the apparatus B was also turned by 90°, it would certainly find that the x-component of spin of the particle B equals $-1/2$. That is reasonable: the total spin must be zero. But, since the apparatus B is still measuring the z-component of spin, it will register either $+1/2$ or $-1/2$, with probability ½.

In the same fashion we can consider the case when the measurement of particle A yields $s_x^A = -\frac{1}{2}$. Then the state of the particle B becomes

$$\frac{|{\downarrow}\rangle + |{\uparrow}\rangle}{\sqrt{2}} = |{\rightarrow}\rangle,$$

and so on and so forth.

---

1   Do not worry about the minus sign. You can multiply an eigenstate by any number, and it remains an eigenstate with the same eigenvalue. In other words: it is the modulus squared of the state vector that matters.

Spotlight: Where a minus does make a difference

What if instead of the EPR state we used the state $|\chi\rangle = \frac{|\uparrow\downarrow\rangle + |\downarrow\uparrow\rangle}{\sqrt{2}}$, in which the expectation value of the z-component of the total spin is also zero (by the way, check this statement)? We can rewrite $|\chi\rangle$ as

$$|\chi\rangle = \frac{1}{\sqrt{2}}\left\{ \frac{|\rightarrow\downarrow\rangle + |\leftarrow\downarrow\rangle}{\sqrt{2}} + \frac{|\rightarrow\uparrow\rangle - |\leftarrow\uparrow\rangle}{\sqrt{2}} \right\}$$

$$= \frac{1}{\sqrt{2}}\left\{ \frac{|\rightarrow\downarrow\rangle + |\rightarrow\uparrow\rangle}{\sqrt{2}} + \frac{|\leftarrow\downarrow\rangle - |\leftarrow\uparrow\rangle}{\sqrt{2}} \right\}.$$

If now we measure the x-component of spin of the particle A, the state of the particle B will be either $\frac{|\downarrow\rangle + |\uparrow\rangle}{\sqrt{2}} = |\rightarrow\rangle$ $\left(\text{if } s_x^A = \frac{1}{2} \right)$ or $\frac{|\downarrow\rangle - |\uparrow\rangle}{\sqrt{2}} = -|\leftarrow\rangle \left(\text{if } s_x^A = -\frac{1}{2} \right)$. In other words, if the system is in the state $|\chi\rangle$, both particles' spins are pointing in the *same* direction. (Please do not think that this means that the expectation value $\langle \chi | \hat{S}_x | \chi \rangle = 1$. It is zero: there will be equal contribution from the cases $s_x^A = s_x^B = \frac{1}{2}$ and $s_x^A = s_x^B = -\frac{1}{2}$.)

As you may recall from Chapter 7, if the total angular moment of a system is $S\hbar$, then its projection on any given axis can take exactly $(2S + 1)$ values – in our case, $2 \times 1 + 1 = 3$. The state $|\chi\rangle$ has $S_z = 0$. The other two states are $|\uparrow\uparrow\rangle$ with $S_z = 1$ and $|\downarrow\downarrow\rangle$ with $S_z = -1$. These three states form a *triplet*, while the EPR state with its zero spin and therefore zero projection of spin on any axis is a *singlet*.

Here is the core of the EPR controversy. The measurement is a physical interaction between a quantum object and an apparatus. This interaction in the point A cannot instantaneously influence the measurement at the point B. Nevertheless, depending on the setup of the detector A and the result of its interaction with the particle A, the detector B, measuring its part (particle B) of the same system in the same initial $|EPR\rangle$- quantum state, will produce either always the opposite of the result of A (if the detectors are aligned), or $\pm\frac{1}{2}$

with equal probabilities (if the detectors are at 90°). What is worse, the detector A can be reset at any moment, e.g. right before the particle A arrives, which completely eliminates any possibility of a signal from A reaching B to tell about the setup. It looks like the events at A and B are, after all, connected by what Einstein called 'spukhafte Fernwirkung' or 'spooky action at a distance'. Einstein was very unhappy about it. According to Einstein, the particles A and B should have real (that is, really existing) values of z- and x-components of their spins all the time.[2] The fact that the operators σ_z and σ_x do not commute and therefore cannot have common eigenstates meant to Einstein that quantum mechanics is not complete: the state vector does not tell us everything about the system.

Niels Bohr responded by saying that the correct way of describing a quantum system depends on how this system interacts with its macroscopic environment (e.g. with the measuring apparatus). For example, a system of two particles can be described by the operators $\hat{p}_1, \hat{x}_1, \hat{p}_2, \hat{x}_2$ of their momenta and positions such that $[\hat{x}_1, \hat{p}_1] = i\hbar, [\hat{x}_2, \hat{p}_2] = i\hbar$, but $[\hat{x}_1, \hat{p}_2] = 0, [\hat{x}_1, \hat{x}_2] = 0$, etc.

> *On this point of view, since either one or the other, but not both simultaneously, of the quantities P and Q can be predicted, they are not simultaneously real. This makes the reality of P and Q depend upon the process of measurement carried out on the first system, which does not disturb the second system in any way. No reasonable definition of reality could be expected to permit this.*
>
> (Einstein et al., 1935: 777)

On the other hand, we can introduce the 'average' and 'relative' quantum operators: $\hat{P} = \frac{\hat{p}_1 + \hat{p}_2}{\sqrt{2}}, \hat{p} = \frac{\hat{p}_1 - \hat{p}_2}{\sqrt{2}}, \hat{Q} = \frac{\hat{x}_1 + \hat{x}_2}{\sqrt{2}}, \hat{q} = \frac{\hat{x}_1 - \hat{x}_2}{\sqrt{2}}$. Obviously (check it!)

---

2   The EPR paper used a different model, but the gist of it is the same.

$[\hat{Q},\hat{P}] = i\hbar, [\hat{q},\hat{p}] = i\hbar$, while $[\hat{Q},\hat{p}] = 0, [\hat{q},\hat{p}] = 0$, etc. We can observe simultaneously $\hat{Q}$ and $\hat{p}$ or, for example, $\hat{p}_1$ and $\hat{q}_2$. Which pair of these can be observed depends on the apparatus (that is, on the macroscopic environment of the system). We may not like the way it all works out, but this is the way it is. It turned out that one cannot reproduce the results of quantum theory with any kind of local hidden variables. The action at a distance is back, though in a spooky way, which Newton probably would not appreciate.

[The] finite interaction between object and measuring agencies conditioned by the very existence of the quantum of action entails – because of the impossibility of controlling the reaction of the object on the measuring instruments if these are to serve their purpose – the necessity of a final renunciation of the classical ideal of causality and a radical revision of our attitude towards the problem of physical reality.

(Bohr, 1935: 696)

[In his response to the EPR argument] Bohr was inconsistent, unclear, wilfully obscure and right. Einstein was consistent, clear, down-to-earth and wrong.

John Bell to Graham Farmelo

Entanglement and faster-than-light communications

The special property of the EPR-state of two particles is that the interaction of one of them with the detector influenced the other even though there was no possibility of a direct physical interaction. To describe such states Erwin Schrödinger introduced the term *entanglement* (*Verschränkung* – at that time the plurality of theoretical physics was still done in German).

We have seen that if you measure the quantum state of the particle A in the entangled state $|EPR\rangle$, the quantum state of the particle B changes depending on the result of this measurement.[3] We can even draw a table illustrating the results of measurements performed on two particles of spin 1/2 in the EPR state, $|EPR\rangle = \frac{|\uparrow\downarrow\rangle - |\downarrow\uparrow\rangle}{\sqrt{2}}$:

Table 11.1 Measurements performed on two particles of spin 1/2 in the EPR state

Measurement at *A*		Probability	State *A* after	State *B* after	Later measurement of S_z^B at *B*	Probability for a given S_z^B
S_z^A	+1/2	50%	$|\uparrow\rangle$	$|\downarrow\rangle$	−1/2	100%
	−1/2	50%	$|\downarrow\rangle$	$|\uparrow\rangle$	+1/2	100%
S_x^A	+1/2	50%	$|\rightarrow\rangle$	$|\leftarrow\rangle$	+1/2	50%
					−1/2	50%
	−1/2	50%	$|\leftarrow\rangle$	$|\rightarrow\rangle$	+1/2	50%
					−1/2	50%

Could it be any different? Yes, of course. Take, for example, the state $|\psi\rangle = |\uparrow\uparrow\rangle = |\uparrow\rangle|\uparrow\rangle$. In this state both particles *A* and *B* have the same spin (+1/2) in *z*-direction. The state vector is a product of states of particle *A* and particle *B*. Therefore whatever measurement is performed on the particle *A*, the state of the particle *B* will not change:

Table 11.2 Measurements performed on two particles of spin 1/2 in *z*-direction

Measurement at *A*		Probability	State *A* after	State *B* after	Later measurement of S_z^B at *B*	Probability
S_z^A	+1/2	100%	$|\uparrow\rangle$	$|\uparrow\rangle$	+1/2	100%
S_x^A	+1/2	50%	$|\rightarrow\rangle$	$|\uparrow\rangle$	+1/2	100%
	−1/2	50%	$|\leftarrow\rangle$	$|\uparrow\rangle$	+1/2	100%

---

3   It is worth repeating here: the words 'measurement' and 'observation' are used as shorthand and in no way imply a conscious effort on anybody's behalf to actually measure anything. The interaction with an appropriate macroscopic physical system is all that matters.

The same is true about the state, where both particles have the same spin in x-direction,

$$|\psi'\rangle = |\rightarrow\rightarrow\rangle = \left(\frac{|\uparrow\rangle + |\downarrow\rangle}{\sqrt{2}}\right)\left(\frac{|\uparrow\rangle + |\downarrow\rangle}{\sqrt{2}}\right) = \frac{|\uparrow\uparrow\rangle + |\downarrow\uparrow\rangle + |\uparrow\downarrow\rangle + |\downarrow\downarrow\rangle}{2},$$

and generally for *any* state, which is a product of functions, describing the particles A and B:

$$|\Psi\rangle = |A\rangle|B\rangle.$$

Such states are called *factorized*. If a system is in a factorized quantum state, one can measure one particle without disturbing the other. Any non-factorized state is *entangled*.

Key idea: Factorized or entangled?

More generally, if a quantum system consists of subsystems A, B, C..., then its quantum state is *factorized* if it can be presented as a product of quantum states of separate subsystems: $|ABC ...\rangle = |A\rangle|B\rangle|C\rangle$ ... Then each of the subsystems can be measured without disturbing the rest.

All non-factorized states are *entangled*, and there are different degrees of entanglement. Examples of *maximally entangled states* are the EPR state and *GHZ* state (after *Greenberger, Horne* and *Zeilinger*, who introduced it in 1989]: $\frac{|\downarrow\downarrow\downarrow\rangle + |\uparrow\uparrow\uparrow\rangle}{\sqrt{2}}$. Such states can be realized in many systems, not necessarily spin-1/2 particles. In experiments on quantum computing and quantum communications one usually uses qubits and photons.

Entangled states make the spooky action at a distance a reality. Their existence underlines that quantum mechanics is not only random, but thoroughly nonlocal. Measuring one particle from an entangled pair will instantaneously change the quantum state of another one, no matter how far it is. Would not this be a marvellous way of beating the special relativity with its measly speed-of-light limit on the speed of communications?

The answer is negative. Suppose we have set up a 'superluminal communication channel' by placing somewhere in space a source of entangled pairs of particles of spin ½ (or photons: instead of spins, we could use the polarization of light, and instead of detectors, polarizers, but that would not change the outcome). One particle of each pair arrives at the detector A, which we will use as a transmitter, and the other (a little later) at the detector B (serving as a receiver). Let every pair be in an EPR state, and assume that none is lost in space, detectors work perfectly, and so on and so forth. If both A and B detectors are aligned in z-direction, then every time there is a +1/2 –readout there, there will be a –1/2-readout at B:

Table 11.3 Results when A and B detectors are aligned in z-direction

A readout: S_z	+	+	–	+	–	...
B readout: S_z	–	–	+	–	+	...

Unfortunately, the +1/2 and –1/2-readouts of s_z at A happen at random, with a probability of 50%, and there is no way of controlling them. The operator at B will be 'receiving' a random string of pluses and minuses (or zeros and ones). Of course, once the operators A and B compare their notes, they will immediately see that they are perfectly anticorrelated. But this is no good for superluminal communications, since the fastest possible way of exchanging these records is by using light or radio waves.

We can try and play with the detector A. Let us, for example, periodically switch it from z- to x-position. Unfortunately, this will only destroy the correlations between the readouts at A and B. You can look at the table of outcomes and see that if at A the x-component is measured, then the z-component at B will be +1/2 or –1/2 with equal probability, whatever is the readout at A.

So, in the end it is still impossible to use the 'action at a distance' to send any information (or, which is the same, physically influence anything[4]) at a superluminal speed.

Spotlight: Aether and EPR

But I was thinking of a plan

To dye one's whiskers green,

And always use so large a fan

That they could not be seen.

This quote from Lewis Carroll's *Through the Looking Glass* was used by Martin Gardner in his classic *Relativity Simply Explained* (with magnificent pictures by Anthony Ravielli) to illustrate early attempts to reconcile the existence of luminiferous aether (a hypothetical all-penetrating medium, in which light waves were supposed to propagate) with the experimentally demonstrated impossibility of observing it. The contrived way one had to go about it eventually led to the development of special relativity, which got rid of aether altogether.

I believe this is an even better illustration of the way, in which quantum randomness and quantum nonlocality contrive *not* to violate special relativity in the EPR experiment.

Spooks acting at a distance, no-cloning and quantum tomography

This does not mean that the EPR effect cannot have its uses in communications. Suppose you want to send a sensitive message through a non-secure line. It must be encrypted, but then the only totally safe encryption method is by using one-time pads. The idea is simple: (i) take a long list of random numbers; (ii) make two copies; (iii) keep one yourself and give the other to your correspondent; (iv) before sending a message, encrypt each letter in turn by using a number from the list; (v) destroy the list (Figure 11.3).

---

4   It is impossible to send information without a material carrier: 'No information without representation' as Rolf Landauer said.

Classical

1-1234	1-1234
2-4678	2-4678
3-9067	3-9067
4-AUG6	4-AUG6
5-77UJ	5-77UJ
6-6531	6-6531
7-HG09	7-HG09
8-7751	8-7751
9-1132	9-1132
10-34FZ	10-34FZ
11-Pq09	11-Pq09

EPR

Figure 11.3 Exchanging one-time pads

The encryption works as follows. Suppose the 13th letter in your message is Y (number 25 in the alphabet), and the 13th number in the one-time pad is 333. You add 25 to 333 modulo 26 (26 because there are 26 letters in Latin alphabet):

$$25 + 333 \ (\text{mod } 26) = 20.$$

The letter number 15 is T, and this letter will replace Y in the message. The recipient, seeing T against the number 333 in his pad, will subtract 333 from 20 modulo 26:

$20 - 333 \pmod{26} = 20 - 333 + 13 \times 26 \pmod{26} = 25$
$\pmod{26} = 25.$

Key idea: Modular arithmetic

Addition modulo Q:

a + b (mod Q) is the remainder after division of (a + b) by Q.
(If a + b <0, one has first to add to a + b such number mQ, that a + b + mQ >0.)

Subtraction modulo Q:

a – b (mod Q) is the remainder after division (a – b) by Q.
(If a – b <0, one has first to add to a – b such number mQ, that a – b + mQ >0.)

This will give the recipient letter Y of the original message. If you like programming and want some privacy, you can write a simple code for encryption and decryption and use one-time pads to exchange secret emails with your friends and family. (The most time-consuming part will be the generation and distribution of these pads.)

The idea behind this encryption method is actually very simple, despite all the 'adding/subtracting modulos' (Figure 11.4). Imagine the alphabet written around a circle. Starting from the given letter, we make as many steps clockwise as the one-time pad tells us to. For example, starting from Y, in 333 steps we arrive at T. When decoding, we make 333 steps counterclockwise starting from T, and return to Y.

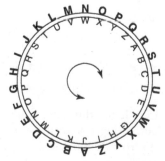

Figure 11.4 Encryption, decryption and addition/subtraction modulo 26

The strength of this kind of encryption is that there are absolutely no statistical correlations in the coded message, which could be used by the cryptanalysts to break the code. Another letter Y in the message (say, on the 227th place) will be encrypted using another, totally unrelated random number, e.g. 616: 25 + 616 (mod 26) = 17; therefore this time Y → Q. So as long as Eve does not have access to the one-time pads, and each pad is only used once and then destroyed, the system is unbreakable.

The only problem is how to securely distribute the one-time pads and ensure that they are really unique and random. And here the EPR pairs can be of assistance (Figure 11.5). Let the sender and the recipient (traditionally called Alice and Bob) align their detectors and receive a stream of EPR pairs. Then each will register a random sequence of zeros and ones (or ups and downs, corresponding to the detected spins). These sequences can be transformed into lists of random binary numbers, that is, ready one-time pads. They *are* as random as one can get, because their randomness follows directly from the fundamental randomness of Nature. They are identical (Alice's zeros correspond to Bob's ones and vice versa – they can agree beforehand, that Bob will flip his sequence). And they cannot be stolen or listened to on the way.

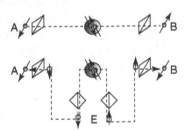

Figure 11.5 A very simple (and not quite practical) scheme for quantum cryptography

The last point deserves some explanation. Suppose an evildoer (traditionally called Eve) can intercept both EPR particles, measure their spins and then send them on towards unsuspecting Alice and Bob (Figure 11.5). But Eve does not know in what direction Alice and Bob aligned their detectors. For example, let her measure the x-components of spins, while Alice and Bob measure their z-components. After Eve's interference the quantum state of the former EPR pair will be either $|\leftarrow\rightarrow\rangle = |\leftarrow\rangle|\rightarrow\rangle$ or $|\rightarrow\leftarrow\rangle = |\rightarrow\rangle|\leftarrow\rangle$. Either of these states is factorized, so their measurements by Alice and Bob will yield the z-component of spin equal to $\pm 1/2$, and there will be no correlation between these outcomes. Eve will corrupt the one-time pads, but will not be able to steal them. Moreover, the appearance of garbage bits will tell Alice and Bob that somebody was eavesdropping. (And this can only be noticed after Alice and Bob exchange information – which can travel no faster than the speed of light.)

Of course, if Eve happens to align her detectors close enough to Oz, she may be able to steal the code with a finite probability (the more the better is the alignment). If, for example, she also measures z-components, she will be sending to Alice and Bob anticorrelated pairs: either $|\uparrow\rangle|\downarrow\rangle$ or $|\downarrow\rangle|\uparrow\rangle$. Alice and Bob will be unable to detect the eavesdropping. But since Alice and Bob can choose any direction at all, and will keep it secret, the chances for Eve to do so are minuscule. This is, of course, the most primitive version of quantum key distribution using entangled states, but the details of better practical schemes are more IT than quantum physics.

Eve's fundamental frustration has everything to do with the basic tenets of quantum mechanics: by measuring a quantum state one changes it. In this particular case the very encryption pad is being created by the very process of measurement of the EPR state carried out by Alice and Bob. But even if the transmitted quantum state *did* carry some useful information, Eve could not extract it covertly nor efficiently. This is prohibited by the *no-cloning theorem*:

It is impossible to copy an unknown quantum state.

To be more specific, consider two quantum systems (for example, qubits). One of them is in a known initial state (call it $|\alpha\rangle$) and another in an unknown quantum state $|\beta\rangle$. Cloning means that the state vector of the system will change from $|\alpha\rangle|\beta\rangle$ to $|\beta\rangle|\beta\rangle$: the states of both qubits are now identical. Then the theorem states that in the general case there is no such Hamiltonian

$\hat{H}_{\alpha\beta} \to _{\beta\beta}$, that an evolution described by the Schrödinger equation $i\hbar\frac{\partial}{\partial t}|\Psi\rangle = \hat{H}_{\alpha\beta\to\beta\beta}(t)|\Psi\rangle$ would take the state of the system from $|\Psi(t=0)\rangle = |\alpha\rangle|\beta\rangle$ to $|\Psi(t=T)\rangle = |\beta\rangle|\beta\rangle$. Therefore Eve cannot make a copy of the state $|\beta\rangle$, keep it for further investigation, and send the original to Alice and/or Bob.

The no-cloning theorem does not say that you cannot reproduce a *known* quantum state. But to know a quantum state means knowing its projections on all the axes of the Hilbert space, that is, measuring it in an appropriate basis as many times as the dimensionality of the Hilbert space. This is what is being done in the process called *quantum tomography* of an unknown quantum state. Since a measurement generally changes the quantum state, you would need at least as many copies of your system in this state, while by the conditions of the theorem you have only one.

Now suppose you do not know the quantum state of your system, but you can repeatedly prepare it in this state. There is no contradiction here. You can always initialize your system in some eigenstate $|\alpha_n\rangle$ of an observable $\hat{A}$ by measuring this observable and using only the outcomes when the measured eigenvalue is a_n. Then you can evolve the system using the same Hamiltonian $\hat{H}(t)$ (e.g. applying a given sequence of pulses of microwave or laser radiation), and the result should be each time the same unknown quantum state vector $|\Psi\rangle$ (allowing for the inevitable experimental errors, of course). Therefore we can measure this state repeatedly and find it. If the Hilbert state of the system is infinite-dimensional, or if it is in a mixed state, the procedure is more involved, but will yield the quantum state of the system with arbitrarily high accuracy (at least, in theory). We will discuss the cost of this accuracy in the next chapter.

Key idea: The VENONA project

No encryption system is *absolutely* unbreakable. Part of the Soviet traffic encrypted using one-time pads was broken in the US-UK VENONA project, when it turned out that some of the pads were accidentally duplicated. It required the statistical analysis of a massive amount of intercepted information accumulated over many years. One-time pads can be lost, stolen or bought. The example of Edward Snowden shows what a single defector can do. Encryption programs can contain backdoors created on purpose by the developers.

If you cannot break a code, break the coder.
Field cryptography 101

Schrödinger's cat and Wigner's friend

If we are going to stick to this damned quantum-jumping, then I regret that I ever had anything to do with quantum theory.
Erwin Schrödinger, as quoted by Heisenberg and requoted in
Van der Merwe 1987: 240

Figure 11.6 Schrödinger's cat Gedankenexperiment (Zagoskin 2011: 4, Fig. 1.1)

The fundamental randomness and nonlocality of quantum mechanics made many physicists unhappy, even though it continued to give extremely accurate predictions in continually expanding areas – from nuclear fission, particle physics and lasers to electronics, materials science and quantum chemistry.

Some of this unhappiness could be attributed to the attitude of some physicists and philosophers, who saw in quantum mechanics the downfall of traditional scientific objectivism, according to which whatever we observe and investigate exists independently of our conscience, and trumpeted the triumph of the subjectivist worldview. The news about the demise of scientific objectivism was premature, though our understanding of what is 'objective reality' had to be seriously amended in view of what quantum mechanics tells us about the world.

The famous 'Schrödinger's cat' Gedankenexperiment is a good example (Figure 11.6). A cat is placed in a closed container together with a radioactive atom (e.g. polonium-210), and a Geiger counter connected to the valve of a can of poisonous gas. Polonium is an α-radioactive element with half-life $T = 138$ days, which decays into the stable isotope, lead-206. The quantum state of the polonium atom can be therefore written as a superposition

$$|\psi(t)\rangle = 2^{-\frac{t}{2T}}\left|{}^{210}_{84}\text{Po}\right\rangle + \sqrt{1 - 2^{-\frac{t}{T}}}\left|{}^{206}_{82}\text{Pb} + \alpha\right\rangle.$$

Measuring the state of the system is to ascertain whether at time t the α-particle escaped the polonium nucleus (with the probability $1 - 2^{-\frac{t}{T}}$) or not (with the probability $2^{-\frac{t}{T}}$). The escape is detected by the Geiger counter, which opens the valve, releases the gas and kills the cat.

The problem is that if we want to be consistent, a state vector should describe the state of the whole system, including the cat:

$$|\Psi(t)\rangle = 2^{-\frac{t}{2T}}\left|{}^{210}_{84}\text{Po}\right\rangle|\text{closed can}\rangle|\text{live cat}\rangle$$
$$+ \sqrt{1 - 2^{-\frac{t}{T}}}\left|{}^{206}_{82}\text{Pb} + \alpha\right\rangle|\text{open can}\rangle|\text{dead cat}\rangle.$$

It is not quite correct to say that the cat is in a quantum superposition of 'live' and 'dead' states: it is the system as a whole that is in such a superposition. (You can see that the quantum states of the cat and the rest of the system are entangled; the state $|\Psi(t)\rangle$ reminds the GHZ state, $\left(|\uparrow\uparrow\uparrow\rangle + |\downarrow\downarrow\downarrow\rangle\right)/\sqrt{2}$)).

But the point is that we never observe the system in such a superposition state: opening the container we will always find either a live or a dead cat. The question is, when and why the quantum state vector collapses (i.e. is projected on either

$$\left|{}^{210}_{84}\text{Po}\right\rangle\left|\text{closed can}\right\rangle\left|\text{live cat}\right\rangle \text{ or } \left|{}^{206}_{82}\text{Pb}+\alpha\right\rangle\left|\text{open can}\right\rangle\left|\text{dead cat}\right\rangle.).$$

An extreme subjectivist point of view states that the collapse is caused by the act of observation by a sentient being (e.g. a human). Logically it cannot be refuted, but not all statements, which cannot be logically refuted, are true.[5] Eugene Wigner proposed an extension of the cat Gedankenexperiment, in which Wigner himself does not open the container. Instead he sends in his friend and waits for his report. Then, the vector should describe the state of the system up to the moment when the friend returns (assuming that Wigner's friend likes cats):

$$\left|\Xi(t)\right\rangle = 2^{-\frac{t}{2T}}\left|{}^{210}_{84}\text{Po}\right\rangle\left|\text{closed can}\right\rangle\left|\text{live cat}\right\rangle\left|\text{happy friend}\right\rangle$$
$$+ \sqrt{1-2^{-\frac{t}{T}}}\left|{}^{206}_{82}\text{Pb}+\alpha\right\rangle\left|\text{open can}\right\rangle\left|\text{dead cat}\right\rangle\left|\text{sad friend}\right\rangle.$$

This state collapses only when Wigner sees his friend and this is not quite a reasonable picture despite its logical irrefutability. Then, there is a question of how sentient the observer must be. Will a cat be able to 'observe itself'? Will there be any difference between a clever and a dumb cat (or a clever and dumb friend)? How about mice? Where do we stop? Will getting a PhD improve your ability to collapse a quantum state?

A more reasonable approach would be to ascribe the collapse to the interactions inside the container, which will either kill or not kill the cat even before anyone bothers to look. The umbrella term for these interactions is *decoherence*. The key requirement

---

5 For example, one can claim that he or she is the only person in the universe and everything else is just an illusion. This is a logically unassailable position – but trying to live it would be a disaster, and writing books promoting such a point of view (solipsism) to allegedly illusory readers is not quite consistent.

of decoherence is the presence of a macroscopic system, but the question remains how macroscopic it should be.

The problems concerning the existence of a macroscopic quantum system in a quantum superposition of states could be considered of no practical interest, until it turned out that quite large systems could be put in a quantum superposition of different states, if only one manages to properly insulate them from the environment. We will therefore consider these questions in the next chapter, when discussing quantum computing.

Bell's inequality*

The nonlocality of quantum mechanics found its exact mathematical expression in Bell's inequality (derived by John Stewart Bell (1928–1990) in 1964).

Bell was interested in what kind of a theory with hidden variables could reproduce the predictions of quantum mechanics, particularly for the EPR experiment.

Let us once more consider two spin-1/2 particles with zero total spin, which are measured by detectors at points A and B, and assume that there is a set of hidden variables Λ, which completely describe these two particles. If we reject action-at-a-distance and insist on locality, then the result of measurement at A must be independent on that at B (Figure 11.7). For example, the probability of measuring spin up at A and spin down at B will be

$$P_{AB}(\uparrow\downarrow;\Lambda) = P_A(\uparrow;\Lambda)P_B(\downarrow;\Lambda).$$

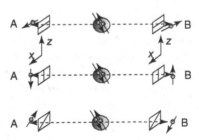

Figure 11.7 Checking Bell's inequality

In this situation the probabilities depend on the values of hidden variables, Λ. Here $P(ab)$ is the probability that both event a and event b occur. You will easily persuade yourself that if these events are independent, then $P(ab)=P(a)P(b)$. For example, the probability of throwing two sixes with fair dice is

$$\frac{1}{6} \times \frac{1}{6} = \frac{1}{36}.$$

In principle a theory with hidden variables can be probabilistic, with the outcome of an experiment only depending on the values Λ through the probability density $\rho(\Lambda)$ But we will show that in order to properly describe the EPR experiment, a theory with local hidden variables must be *deterministic*: for each state of the system there must be a definite choice of Λ.

Indeed, for an EPR pair of particles the spins, measured at A and B, are always opposite. Therefore

$$P_{AB}(\uparrow\uparrow) = 0 = \int d\Lambda \, P_{AB}(\uparrow\uparrow; \Lambda)\rho(\Lambda) = \int d\Lambda \, P_A(\uparrow; \Lambda) \, P_B(\uparrow; \Lambda)\rho(\Lambda).$$

Probabilities and probability densities are non-negative. Therefore the integral on the right-hand side can be zero only if either $P_A(\uparrow; \Lambda) = 0$, or $P_B(\uparrow; \Lambda) = 0$, or $\rho(\Lambda) = 0$. The last choice is nonsensical (it would mean that nothing ever happens, including 'nothing'). Therefore we are left with either $P_A(\uparrow; \Lambda) = 0$, or $P_B(\uparrow; \Lambda) = 0$. In the same way from $P_{AB}(\downarrow\downarrow) = 0$ we find that either $P_A(\downarrow; \Lambda) = 0$, or $P_B(\downarrow; \Lambda) = 0$. Each measurement can produce either spin up, or spin down. Therefore the result of the measurement is determined by the hidden variables without any randomness: if $P_A(\uparrow; \Lambda) = 0$, then $P_A(\downarrow; \Lambda) = 1$, and so on. This is already remarkable: locality demands determinism.

Now let us conduct *three* different measurements of an EPR pair at A and B. For example, we can measure s_x, s_z and the projection of spin on the diagonal at $45°$ in the plane Oxz. The measured spins must be always opposite. In a theory with hidden parameters, the outcomes of these experiments must be determined by some pre-existing (before the measurement) variables. Let us call ξ_A the variable, which determines the outcome of the s_x measurement on particle A: $\xi_A = \pm 1$ if the measurement gives $s_x^A = \pm\frac{1}{2}$. In the same way we introduce $\zeta_A = \pm 1$ for the measurement of s_z^A, and $\eta_A = \pm 1$ for the measurement of $s_{45°}^A$, and similar variables for the measurements performed on the particle B.

Now, the probability for a particle from the EPR pair to have values $\zeta = 1, \eta = -1$ equals

$$P(\zeta = 1, \eta = -1) = P(\xi = 1, \zeta = 1, \eta = -1) + P(\xi = -1, \zeta = 1, \eta = -1)$$
(since either $\xi = 1$, or $\xi = -1$).

Probabilities are non-negative, therefore

$$P(\xi = 1, \zeta = 1, \eta = -1) \le P(\xi = 1, \zeta = 1) \text{ and}$$
$$P(\xi = -1, \zeta = 1, \eta = -1) \le P(\xi = -1, \eta = -1), \text{ and we have}$$

$$P(\zeta = 1, \eta = -1) \le P(\xi = 1, \zeta = 1) + P(\xi = -1, \eta = -1).$$

For an EPR pair $\xi_A = -\xi_B$ etc. Therefore the last line can be rewritten as *Bell's inequality* for the outcome probabilities of measurements on *two* EPR particles:

$$P(\zeta_A = 1, \eta_B = 1) \le P(\xi_A = 1, \zeta_B = -1) + P(\xi_A = -1, \eta_B = 1).$$

Any theory with local hidden variables must satisfy this inequality. On the other hand, one can show that in quantum theory this inequality can be violated under certain conditions. Therefore an experimental observation of the violation of Bell's inequality is proof positive of nonlocality. (If Bell's inequality is satisfied in an experiment, it does not mean that the quantum theory is wrong – just that the experiment is not stringent enough).

By now the violation of Bell's inequality was observed a number of times, in experiments with electron–positron pairs, with protons, with photons and even with qubits. So there is no doubt: whatever it is, Nature is not local.

Fact-check

1 The EPR experiment demonstrates that quantum mechanics is

   a stochastic

   b nonlocal

   c acausal

   d incomplete

2 The EPR experiment does not allow faster-than-light communications, because

   a the wave function collapse takes finite time

   b the outcome of a quantum measurement can only be predicted statistically

   c the measurements at A and B are not correlated

   d the total spin is only conserved on average

3 Which of these quantum states are entangled?

   a $\dfrac{|\uparrow\uparrow\rangle + |\downarrow\uparrow\rangle - |\uparrow\downarrow\rangle - |\downarrow\downarrow\rangle}{2}$

   c $\dfrac{3|\uparrow\downarrow\rangle - |\downarrow\downarrow\rangle}{\sqrt{10}}$

   b $\dfrac{|\uparrow\uparrow\rangle - |\downarrow\downarrow\rangle + 2|\uparrow\downarrow\rangle}{\sqrt{6}}$

   d $\dfrac{3|\downarrow\uparrow\rangle + |\uparrow\downarrow\rangle}{\sqrt{2}}$

4 What are some of the vulnerabilities of one-time pad cryptography?

   a repeated use

   b secure distribution

   c password stealing

   d redundancy

5 What are some of the advantages of quantum key distribution?

   a negligible chance of eavesdropping

   b faster-than-light communication

   c one-time pads of arbitrary length

   d eavesdropping attempts can be detected

6 According to the no-cloning theorem

   a no two quantum systems can be prepared in the same quantum state

   b no quantum system can be prepared in a superposition of two quantum states

c an unknown quantum state cannot be copied
d a known quantum state cannot be copied

7 The 'Schrödinger's cat' and 'Wigner's friend'
Gedankenexperiments illuminate

a the quantum measurement problem
b the problem of quantum-classical transition
c the influence of the observer's conscience on quantum
mechanical processes
d the incompleteness of quantum mechanics

8 An experiment, in which the Bell's inequality is violated,
demonstrates that

a quantum mechanics is incomplete
b quantum mechanics is self-contradictory
c local hidden variables are possible
d Nature is non-local

9 The macroscopic environment of q quantum system

a disrupts quantum correlations in the system
b makes quantum evolution impossible
c reduces the dimensionality of the system's Hilbert space
d determines which observables can be measured

10 In an EPR experiment, what is the probability of measuring
spin up at A *and* spin down at B?

a 100%
b 50%
c 25%
d 0%

Dig deeper

M. Le Bellac, *A Short Introduction to Quantum Information and Quantum Computation*. Cambridge University Press, 2007 (Chapters 2 and 4).

A discussion of entanglement is in Chapter 23 of

R. Penrose, *The Road to Reality: A Complete Guide to the Laws of the Universe*. Vintage Books, 2005. Starting from there, you will likely want to keep reading this very interesting book in both directions!

The history and basic science of cryptography are described in

S. Singh, *The Code Book*. Fourth Estate, 1999.

The VENONA project is described in detail in

C. Andrew, *The Defence of the Realm: The Authorized History of MI5*. Allen Lane, 2009, p. 367.

12

Schrödinger's elephants and quantum slide rules

The key feature of science is its ability to not just explain, but to *predict* the behaviour of the world around us. From the predictions of eclipses to the calculations of planetary orbits and trajectories of spacecraft and ballistic missiles, from atomic transitions to chemical reactions, from pure science to applied science to engineering – the power of science is built on its predictive force, and the latter means a lot of computations. The reason is that while the mathematical expression of the laws of Nature may be quite simple, the corresponding equations of motion practically never have exact solutions. One must develop either some approximate schemes, or numerical methods of solving these equations, or both. And in any case solving a problem 'from the first principle' remains a very challenging task. In dealing with it we come to rely more and more on the support of computers. But the use of computers creates problems of its own, from developing a general theory of their operation to the practical questions of their design, manufacture and maintenance, to the point when engineering and fundamental physics are no longer distinguishable. This makes the field of quantum computing and – more broadly – quantum engineering especially interesting.

Celestial mechanics – the application of the Newtonian dynamics to the motion of the celestial bodies – was the first science where large-scale computations were required.

The ideal picture of planets following the elliptic orbits around the Sun in accordance with Kepler's laws is easily derived from the Newton's laws and universal gravity, but only if there are only two bodies in the system – the Sun and a single planet. This is just an approximation to reality: there are more than two objects in this Solar system, and, unfortunately it is impossible to solve exactly the equations of motion already for three gravitating objects (except for some very special cases). This is the famous 'three-body problem'. It does have some equally famous exact solutions (some of them found by Lagrange), but it has rigorously proven that this problem has *no* 'good' solutions for the absolute majority of initial conditions.

The situation with more than three objects is even worse. Therefore in order to solve the Newtonian equations of motion for the Solar system (and thereby confirm – or refute – the validity of Newtonian dynamics and universal gravity, by comparing these theoretical predictions with the actual astronomical observations), it was necessary to build the *perturbation theory*. For example, when making calculations for Venus, first one had to take into account the pull from the Sun (which is the greatest), then from Jupiter, then from other planets, and so on and so forth.

In perturbation theory the solutions are found as a sum of many (in principle – infinitely many) contributions, finding which is a time-consuming and labour-intensive process, requiring many computations. First this was done manually (with the help of the logarithmic tables, invented in the 17th century by John Napier). The accuracy of these calculations by the mid-19th century was sufficient to predict and then discover a new planet, Neptune, based on the deviations of Uranus from its calculated orbit. Nevertheless many effects in celestial mechanics – such as the relation between the satellites of giant planets and the shape of their rings – could only be investigated with the arrival of powerful computers in the latter half of the 20th century, because they involved such a large number of bodies that any manual computations, even assisted by mechanical calculators or slide rules, were hopeless.

The growing power of computers prompted the development of a separate branch of physics, computational physics, along with theoretical and experimental physics, devoted to a direct modelling of the behaviour of various physical systems by solving the corresponding equations of motion numerically.[1] It was not only better numerical algorithms that were being developed. The hardware was evolving fast: the speed of computers steadily increased, and their size shrank. According to the famous *Moore's law* – an observation made back in 1965 – the number of transistors in an integrated circuit doubled approximately every two years, and the trend, surprisingly, holds even now. But this shrinking of the unit elements cannot go on forever. There is a limit to *how small* they can become. In principle, this would limit the power of computers, at least those built on the same principles as the current ones. From here followed another important question: whether a computer can *efficiently* model physical systems.

'Efficiently' here means that there should be a reasonable relation between the scale of the problem and the amount of computational resources needed to solve it. If, for example, in order to solve the problem of three gravitating bodies you would need a desktop, for six bodies a supercomputer, and for nine bodies all the computing power of Earth, the modelling would not be efficient. Usually one expects from 'efficiency' that the computational resources should scale no faster than some power of the size of the problem. So an approach requiring one desktop for a three-body problem and 2, $2^2 = 4$ or even, say, $2^{10} = 1024$ desktops for a six-body problem could be considered

---

1   For example, by 'slicing' time and space in small pieces and replacing the differential equations by the difference equations, as we have briefly mentioned in Chapter 2.

'efficient'. (The latter case would not be *really* efficient, but that is a different story.)

Richard Feynman was one of those scientists who were interested in both problems, and in 1982 he pointed out that a computer, which operates according to the rules of classical physics, cannot efficiently model a quantum system. That was an important, troubling and promising observation.

The number of transistors in an integrated circuit doubles approximately every two years.
Moore's Law

If something cannot go on forever, it will stop.
Herbert Stein's Law

The need and promise of quantum computers

Why cannot a classical computer efficiently model a quantum system? In short, the Hilbert space, in which the state vector of a quantum system 'lives', is significantly spacier than the phase space, in which the state of a classical system evolves. Recall, that even for a single particle in one-dimensional well (Chapter 2) the Hilbert space has infinitely many dimensions, while the phase space of a classical particle would have only two dimensions. Another way of looking at the situation is through the Feynman's path integrals. When modelling a classical particle's motion from A to B, we had to deal with a single trajectory. For a quantum particle we would have to take into account *all* possible trajectories. While for one or few quantum particles the equations of quantum mechanics can be solved easily enough, as their numbers grow the task becomes impossible very fast. Of course, for large collectives of quantum particles certain very useful approximate approaches can be found (as we have seen in Chapter 10), but they do not work for *all* possible quantum states. If the quantum state of a many-body system is factorized, for example, $|\uparrow\uparrow\ldots\uparrow\rangle = |\uparrow\rangle_1|\uparrow\rangle_2\ldots|\uparrow\rangle_N$, there will be no problem. (We denote here by '1', '2', etc. the set of variables that describes a single particle). On the other hand, for entangled states which involve many particles, like the maximally entangled states of Chapter 11, say, $\left(|\uparrow\uparrow\ldots\uparrow\rangle - |\downarrow\downarrow\ldots\downarrow\rangle\right)/\sqrt{2}$, these approaches do not work well if at all. In the Hilbert space of a many-body system there are unimaginably more entangled than factorized quantum states, and even if we start from a factorized state, quantum evolution will immediately take us to some entangled state (unless certain very special conditions are met).

To better see the difficulty and a possible way out of it, let us recall that computers are not necessarily something that sits on your desktop or in your bag and uses 'Windows' or another operating system. These are *digital* computers, which encode a problem into a series of numbers and perform step-by-step operations on them, and they are not the only ones possible. We may call a classical computer *any* physical system, operating following the laws of classical physics, which can be used to simulate (and therefore predict) the behaviour of another physical system, be it classical or quantum.

One important class of computers are *analogue* computers. There the problem is encoded not in a series of numbers, but directly in the physical state of the computer, and then this state is allowed to evolve – according to the laws of classical physics – to the state, which encodes the solution. Before digital computers became really powerful, the analogue computers held sway in many areas, where efficient and fast solution of complex problems was necessary, maybe at the expense of precision.

A simple example of an analogue computer is a slide rule (Figure 12.1). It uses the mathematical fact that $\log(ab) = \log a + \log b$ and the physical fact that the distance between any two points of a rigid body is constant. Therefore one can easily find products and ratios, as well as trigonometric functions, powers, roots and a lot of other things (there were specialized slide rules for electrical engineers, skippers, gunners,

Figure 12.1 A slide rule

and so on). Slide rules were fast, efficient and not too expensive, and for a hundred years there was no engineer and hardly a physicist without a slide rule.[2]

More complex analogue computers were built for solving a specific problem, or a class of problems. One area, where one early on had to solve complex equations as accurately as possible, is the continuum mechanics (which includes such disciplines as fluid mechanics and theory of elasticity). Its development was spurred by the requirements of practical engineering. Unfortunately, the equations, which describe such systems – be it an iron strut under varying stress or an airplane wing in a turbulent flow – are notoriously hard to solve. Take, for example, the fundamental equations of fluid mechanics, the *Navier–Stokes equations*, which describe the motion of a viscous fluid and are known for almost 200 years. Not only are they very hard to solve numerically. The question of whether there exists or not a 'smooth and globally defined' solution to these equations remains a major unsolved problem of mathematical physics. But finding accurate enough solutions to these equations for a variety of conditions is absolutely necessary when designing ships and aircraft.

One can use wind tunnels to directly recreate the desired conditions and simply observe what the aircraft will do under such and such conditions. But for a large aircraft or a ship this would not do. It was proven though (back in the 19th century) that a scale model of a ship can be tested at a different speed, and maybe in a different kind of medium, and the results can be recalculated back to the full scale. The installations where such

---

2   In Hayao Miyazaki's last animated film, The Wind Rises, the hero's slide rule plays a prominent part.

tests are performed (*ship model basins*) are essentially analogue computers (or simulators), which use one physical system to predict the behaviour of a different physical system.

When building analogue computers one makes use of the fact that *different* physical systems are often described by *identical* equations. For example, a mechanical oscillator (a spring pendulum with friction) and an electrical oscillator (an LC circuit with resistive loss) satisfy the same equation, $\frac{d^2q}{dt^2} = -\omega^2 q - \eta \frac{dq}{dt}$, even though in one case q is the position of a pendulum and in the other it is voltage on a capacitor, and η is determined by the friction or electrical conductance. Therefore in order to solve a system of differential equations one can build an appropriate electric circuit (which will be much easier to make and measure than an equivalent system of pendulums!) and measure the appropriate voltages.

Taking up this approach, we therefore expect (with Feynman), that the use of *quantum computers* to simulate quantum systems should save us from the tight place. Again, we will call a quantum computer *any* quantum system that can be used to simulate (and therefore predict) the behaviour of another quantum system.

Digital computers, circuits and gates

The major disadvantage of analogue computers is their limited accuracy, limited scalability and non-universality. Their accuracy is indeed limited by material restrictions (e.g. the length and rigidity of a slide rule). Scaling up an analogue computer is a self-limiting task: one can end up with testing the actual system (like an aircraft in the wind tunnel), which is then no longer 'modelling'. The non-universality is also clear: one has to build an analogue computer for a specific task (or a class of similar tasks), and it is not a given that a convenient analogue system for a given task actually exists.

Spotlight: The power of the analogies

Analogue approaches can be surprisingly efficient. For example, in order to sort a random list of N numbers, a digital computer would in the worst case scenario require N^2 operations. Alexander Dewdney in 1984 proposed the 'spaghetti sort', which only takes $2N+1$ operations. The method is simple. (i) Take N uncooked spaghetti rods and cut them to the lengths, which correspond to the numbers from your list (N operations). (ii) Take the rods in your fist, put them down on a horizontal surface and let them slide (one operation): this is where the actual sorting takes place. (iii) List the rods from the tallest to the lowest (N operations).

Key idea: Logic gates

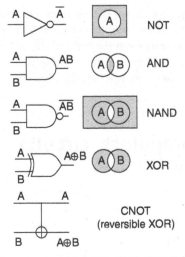

Examples of logic gates include NOT, AND, XOR, NAND and CNOT. The bits (or qubits) are represented by solid lines, which in a representation of an algorithm as a circuit run from the left (input) to the right (output). The Venn circles illustrate the action of the gates: the shaded area corresponds to the value 1 ('TRUE') of the output bit.

Digital computers avoid these difficulties by reducing the solution of a problem to purely symbolic manipulations. Of course, in the final analysis these manipulations are implemented by *physical* processes in some *physical* systems ('no information without representation!'), but these elementary operations are totally independent of the problem to be solved and can be implemented in a universal hardware. The idea of a universal digital computer is an old one, going at least as far back as Pascal and Leibniz. It was almost realized by Charles Babbage,[3] but had to wait for the electron valves and semiconductor transistors.[4]

Taken in the most abstract way, a classical digital computer transforms an *input* into an *output*. An input and an output can be represented as binary numbers, i.e. sequences of *bits*, each taking value zero or one. A computer performs on the bits a series of operations (*logic gates*), which form a *circuit*. If the output is, e.g. the numerical value of some function, there is no difficulty (in principle) to get a more accurate result by increasing the number of bits in the input and registers.

A logic gate takes one or more inputs and transforms them into one or more outputs. Examples are NOT, AND, NAND and XOR, described by the following *truth tables* (coming from logic: when a bit represents the validity of some statement, the value 1 is called TRUE and the value 0 FALSE respectively):

Table 12.1 Truth table 1 (NOT A)

IN	OUT
A	NOT A
0	1
1	0

---

3   William Gibson's and Bruce Sterling's *The Difference Engine* is a classic steampunk novel exploring the consequences of a successful realization of Babbage's computer in the 19th century.

4   As in so many cases, this major development in science and technology was greatly accelerated by the military need (to promptly decipher German military codes produced by the Enigma cipher machine), but the fascinating details are immaterial here.

Table 12.2 Truth table 2 (A AND B)

IN	IN	OUT
A	B	A AND B
0	0	0
0	1	0
1	0	0
1	1	1

Table 12.3 Truth table 3 (A NAND B)

IN	IN	OUT
A	B	A NAND B
0	0	1
0	1	1
1	0	1
1	1	0
IN	IN	OUT
A	B	A XOR B
0	0	0
0	1	1
1	0	1
1	1	0

The first of these gates is *reversible*: you can run it backwards and return to the initial state of the bit register.[5] The other three are obviously *irreversible*: for example, if you only know that (A XOR B) = 1, you cannot tell whether A = 1, B = 0 or the other way around. In principle, they can be made reversible simply by retaining one of the inputs along with the output. The reversible version of XOR is called CNOT ('controlled NOT').

Indeed, we see from the truth table that if the bit A ('control') is in state 0, the state of the bit B ('target') does not change; if A is in state 1, then B flips (that is, undergoes the NOT operation).

Though there are many different logic gates, any circuit (that is, any digital computation) can be performed using

---

5 Any logical circuit can be implemented with reversible gates only, though this is not how conventional computers operate.

Table 12.4 Truth table 4 (A CNOT B)

IN	IN	OUT	OUT
A	B	A = A	B = A XOR B
0	0	0	0
0	1	0	1
1	0	1	1
1	1	1	0

a smaller number of *universal* gates. In classical case, this 'smaller number' is one: any circuit can be implemented using only NAND or only NOR gates (which is convenient for a theoretical analysis).

A *quantum* digital computer manipulates a set of quantum bits (*qubits*) – two-state quantum systems we have investigated in Chapter 4. The state of a qubit is determined not by a single number (1 or 0), but by the quantum state vector in a two-dimensional Hilbert space (which, as we know, is conveniently represented by a Bloch vector). For a set of two qubits, the Hilbert space is four-dimensional, for N qubits it is 2^N-dimensional – which means that while the state of an N-bit classical register is described by just N numbers (zeros or units), the state of an N-qubit register is determined by 2^N complex numbers, or 2^{N+1} real numbers (actually, a little less if we take into account the normalization conditions). This is a huge difference, which illustrates the difficulty of modelling quantum systems with classical means.

Fortunately, at least the manipulations (*quantum gates*), which should be performed on qubits in order to realize an arbitrary operation of a quantum computer, are not very demanding. The *universal set* can be composed of just two kinds of gates.

Quantum gates

First, the set of universal quantum gates includes one-qubit gates, which should allow arbitrary manipulations with the quantum state of a single qubit (just imagine rotating its Bloch vector). Examples of such gates are *Pauli-X, Y and Z* gates, given by the Pauli matrices $\sigma_{x,y,z}$ from Chapter 4. If a qubit is in the state $|0\rangle = \begin{pmatrix} 1 \\ 0 \end{pmatrix}$, then acting by the Pauli matrices on this state we will get respectively

$$X|0\rangle = \sigma_x|0\rangle = \begin{pmatrix} 0 & 1 \\ 1 & 0 \end{pmatrix} \begin{pmatrix} 1 \\ 0 \end{pmatrix} = \begin{pmatrix} 0 \\ 1 \end{pmatrix} = |1\rangle; \quad Y|0\rangle$$

$$= \sigma_y|0\rangle = \begin{pmatrix} 0 & -i \\ i & 0 \end{pmatrix} \begin{pmatrix} 1 \\ 0 \end{pmatrix} = \begin{pmatrix} 0 \\ i \end{pmatrix}$$

$$= i|1\rangle; \quad Z|0\rangle = \sigma_z|0\rangle = \begin{pmatrix} 1 & 0 \\ 0 & -1 \end{pmatrix} \begin{pmatrix} 1 \\ 0 \end{pmatrix} = \begin{pmatrix} 1 \\ 0 \end{pmatrix} |0\rangle.$$

These gates can be applied, of course, to a qubit in an arbitrary state $|\psi\rangle = \begin{pmatrix} u \\ v \end{pmatrix}$ (where $|u|^2 + |v|^2 = 1$).

A useful exercise

Find the results of action by the Pauli gates on qubit states

$$|1\rangle = \begin{pmatrix} 0 \\ 1 \end{pmatrix}; |\Psi_+\rangle = \frac{|0\rangle + |1\rangle}{\sqrt{2}} = \frac{1}{\sqrt{2}} \begin{pmatrix} 1 \\ 1 \end{pmatrix} \text{ and } |\Psi_-\rangle = \frac{|0\rangle - |1\rangle}{\sqrt{2}} = \frac{1}{\sqrt{2}} \begin{pmatrix} 1 \\ -1 \end{pmatrix}.$$

There remains a practical question of *how* does one apply gates to a qubit? Applying a gate means changing the quantum state of a qubit, and quantum states change because they evolve according to the Schrödinger equation: $i\hbar \dfrac{d|\Psi\rangle}{dt} = \hat{H}|\Psi\rangle$.
For a small time interval Δt this means that approximately $|\Psi(t + \Delta t)\rangle - |\Psi(t)\rangle \approx -\dfrac{i\Delta t}{\hbar} \hat{H}|\Psi(t)\rangle$. For many realistic qubits the Hamiltonian is, like in Chapter 4, $\hat{H} = -\frac{1}{2} \begin{pmatrix} \varepsilon & \Delta \\ \Delta & -\varepsilon \end{pmatrix} = -\frac{1}{2}\varepsilon\sigma_z - \frac{1}{2}\Delta\sigma_x$, and the parameters ε and Δ can be changed by literally turning a knob. One can, for example, make $\epsilon = 0$, and this leaves only a term proportional to σ_x applied to the quantum state vector $|\Psi(t)\rangle$.

We get a more clear picture of how to apply a one-qubit gate if we represent the qubit quantum state as a Bloch vector (Chapter 4). There we stated that by playing with the qubit Hamiltonian (that is, turning some knobs) we can turn the Bloch vector from any initial to any final position on the Bloch sphere. This is not an instantaneous process: rotation of the Bloch vector takes time depending on the kind of qubit we use, and there are clever sequences of turns (that is, of changes to the qubit Hamiltonian by turning some knobs), which minimize the required time while maintaining the accuracy of the operation.

Another very useful one-qubit gate is the *Hadamard*[6] *gate*,
$H = \frac{1}{\sqrt{2}} \begin{pmatrix} 1 & 1 \\ 1 & -1 \end{pmatrix}$. As you can see, $H|0\rangle = \frac{|0\rangle + |1\rangle}{\sqrt{2}}$; $H|1\rangle = \frac{|0\rangle - |1\rangle}{\sqrt{2}}$.
In other words, the Hadamard gate puts a qubit, which was initially in a state 'zero' or 'one' (as a classical bit in a classical computer), in a quantum superposition of these two states.

There is actually a wide choice of one-qubit operations, which can be included in the universal set. Which are best to use depends on the particular kind of qubits.

In addition to one-qubit gates one needs one of the *universal two-qubit* gates. Probably the most popular choice is the quantum CNOT gate (Figure 12.2). It operates on two qubits: the control and the target. Like its classical counterpart, it does nothing to the control qubit state, and flips (that is, applies

Figure 12.2 Quantum CNOT gate (Zagoskin 2011: 314, Fig. A.1.1)

---

6   Named after Jacques Hadamard, a prominent French mathematician, and pronounced 'ah-dah-mAHr'.

Pauli-X-gate) to the target qubit state, if the control qubit is in state $|1\rangle$ This gate can be represented as a four-by-four matrix

$$CNOT = \begin{pmatrix} 1 & 0 & 0 & 0 \\ 0 & 1 & 0 & 0 \\ 0 & 0 & 0 & 1 \\ 0 & 0 & 1 & 0 \end{pmatrix}.$$

Indeed, if the control qubit is in state $|0\rangle = \begin{pmatrix} 1 \\ 0 \end{pmatrix}$ and the target qubit is in some state $|\psi\rangle = \begin{pmatrix} u \\ v \end{pmatrix}$, then the two-qubit state vector is

$$|\psi\rangle = \begin{pmatrix} 1 & . & u \\ 1 & . & v \\ 0 & . & u \\ 0 & . & v \end{pmatrix} \begin{pmatrix} u \\ v \\ 0 \\ 0 \end{pmatrix},$$

and applying to it the CNOT gate we obtain

$$CNOT|\psi\rangle = \begin{pmatrix} 1 & 0 & 0 & 0 \\ 0 & 1 & 0 & 0 \\ 0 & 0 & 0 & 1 \\ 0 & 0 & 1 & 0 \end{pmatrix} \begin{pmatrix} u \\ v \\ 0 \\ 0 \end{pmatrix} = \begin{pmatrix} u \\ v \\ 0 \\ 0 \end{pmatrix} = |\psi\rangle,$$

as expected: neither control nor target qubit states change. On the contrary, if the control qubit is in state $|1\rangle = \begin{pmatrix} 0 \\ 1 \end{pmatrix}$, then

$$|\psi'\rangle = \begin{pmatrix} 0 & . & u \\ 0 & . & v \\ 1 & . & u \\ 1 & . & v \end{pmatrix} \begin{pmatrix} 0 \\ 0 \\ u \\ v \end{pmatrix},$$

and the CNOT gate will flip the target qubit:

$$CNOT|\psi'\rangle = \begin{pmatrix} 1 & 0 & 0 & 0 \\ 0 & 1 & 0 & 0 \\ 0 & 0 & 0 & 1 \\ 0 & 0 & 1 & 0 \end{pmatrix} \begin{pmatrix} 0 \\ 0 \\ u \\ v \end{pmatrix} = \begin{pmatrix} 0 \\ 0 \\ v \\ u \end{pmatrix} = |\tilde{\psi}'\rangle (u \leftrightarrow v).$$

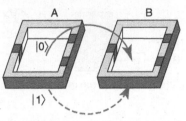

Figure 12.3 Two flux qubits interact via their magnetic fields

There is a rigorous theorem, which states that *any computation on a digital quantum computer can be performed using only universal one-qubit and two-qubit gates* (for example, one-qubit gates and CNOT). This is actually great news. Two-qubit gates require interactions between two qubits, which are relatively easy to realize. For example, two adjacent flux qubits will interact, because the magnetic field of qubit A acting on qubit B depends on the state of qubit A (Figure 12.3). Three- or multi-qubit gates would be much harder to realize.

Quantum parallelism and the Deutsch algorithm*

A quantum system is hard to model by classical means, because in a sense it takes all possible paths from the initial to the final state at once. Loosely speaking, while a classical computer calculates one correct answer, a quantum computer obtains *all* answers at once (mostly the wrong ones), but in the end a clever quantum algorithm makes the wrong answers cancel each other. This is the *quantum parallelism*, which can be illustrated on a simple, but important, example of a quantum algorithm, the *Deutsch algorithm.*[7]

---

7   David Deutsch is a British physicist and one of the founders of quantum computing.

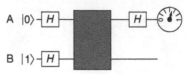

Figure 12.4 The Deutsch algorithm run on a two-qubit quantum computer

Suppose we have an unknown function $f(x)$, which can only take values zero and one, and we want to know whether $f(0) = f(1)$ (such function is called 'constant') or not (then it is called 'balanced'). The values of this function are computed by some 'black box' from the input argument):

$$0 \to f(0); 1 \to f(1).$$

For the reasons that will become clear in a moment, it is more convenient to use a black box, with two inputs and two outputs (Figure 12.4). Let the input A be x (zero or one). Then the modified black box will compute $f(x)$ if the input B is 0 and $(1-f(x))$ if the input B is 1 and send it to the output B:

Table 12.5 Results of modified black box

INPUT A	INPUT B	OUTPUT A	OUTPUT B
x (0 or 1)	0	x	$f(x)$
x (0 or 1)	1	x	$1-f(x)$

If we have a classical computer, there is no choice but to actually calculate and compare $f(0)$ and $f(1)$, because a classical bit can be either zero or one. On the other hand, a quantum bit can be in a superposition of both states, and the Deutsch algorithm takes advantage of this.

Key idea: Heads or tails?

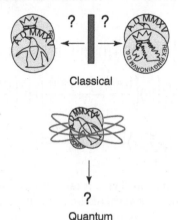

Classical

?

Quantum

The problem solved by the Deutsch algorithm (in its simplest version) is equivalent to checking whether a coin is fair ('balanced') or not ('constant'). Classically, one must look at both sides and compare them, which takes two checks. In the quantum case, the coin is put in a superposition of 'heads' and 'tails', and one look is enough.

In its simplest form it requires a quantum computer with two qubits, A and B (Figure 12.4). We will initialize them in states $|0\rangle$ and $|1\rangle$, and then apply to both the Hadamard gates. Then the quantum state of the two-qubit computer is

$$|\psi\rangle_{in} = \left(\frac{|0\rangle+|1\rangle}{\sqrt{2}}\right)\left(\frac{|0\rangle-|1\rangle}{\sqrt{2}}\right) = \frac{1}{2}\left(|0\rangle|0\rangle+|1\rangle|0\rangle-|0\rangle|1\rangle-|1\rangle|1\rangle\right).$$

Now let us feed these qubit states into the black box. It will change the state of qubit A according to the state of qubit B, that is

$$|\psi\rangle_{in} \to |\psi\rangle_{out} = \frac{1}{2}\left(|0\rangle|f(0)\rangle + |1\rangle|f(1)\rangle - |0\rangle|1-f(0)\rangle - |1\rangle|1-f(1)\rangle\right)$$

$$= \frac{1}{\sqrt{2}}\left(|0\rangle\left(\frac{|f(0)\rangle - |1-f(0)\rangle}{\sqrt{2}}\right) + |1\rangle\left(\frac{|f(1)\rangle - |1-f(1)\rangle}{\sqrt{2}}\right)\right)$$

$$= \frac{1}{\sqrt{2}}\sum_{x=0}^{1}|x\rangle\frac{|f(x)\rangle - |1-f(x)\rangle}{\sqrt{2}}.$$

Recall that the function $f(x)$ can take only values zero and one. Therefore the state of the qubit B in the output, $\frac{|f(x)\rangle - |1-f(x)\rangle}{\sqrt{2}}$, is either $\frac{|0\rangle - |1\rangle}{\sqrt{2}}$ (if $f(x) = 0$) or $\frac{|1\rangle - |0\rangle}{\sqrt{2}} = -\frac{|0\rangle - |1\rangle}{\sqrt{2}}$ (if $f(x) = 1$). We can rewrite the state of the two-qubit system as $|\psi\rangle_{out} = \left\{\frac{1}{\sqrt{2}}\sum_{x=0}^{1}(\pm|x\rangle)\right\}\left\{\frac{|0\rangle - |1\rangle}{\sqrt{2}}\right\}$. The plus sign in '$\pm$' is taken if $f(x) = 0$, minus if $f(x) = 1$. (This sensitivity to $f(x)$ is due to our earlier insistence that the black box should calculate $f(x)$ or $(1-f(x))$ depending on the state of the qubit B).

Note also that the quantum state $|\psi\rangle_{out}$ turns out to be factorized: it is the product of the state of qubit A and the state of qubit B. We can therefore now forget about the qubit B altogether (and anyway its state, $\frac{|0\rangle - |1\rangle}{\sqrt{2}}$, does not depend on $f(x)$) and concentrate our attention on qubit A.

We wanted to know whether $f(0) = f(1)$ or not. If $f(0) = f(1)$, then the output state of the qubit A is $\pm\frac{|0\rangle + |1\rangle}{\sqrt{2}} = \pm|\psi_+\rangle$. If $f(0) \neq f(1)$, then the state is $\pm\frac{|0\rangle - |1\rangle}{\sqrt{2}} = \pm|\psi_-\rangle$. The overall sign of the state vector does not matter, therefore we can say that if the qubit A is found in the state $|\Psi_+\rangle$, the function $f(x)$ is constant; if it is in the state $|\Psi_-\rangle$, $f(x)$ is balanced.

It is usually convenient to measure a qubit in one basis of states $\{|0\rangle, |1\rangle\}$, and in this basis we cannot tell the state $\frac{|0\rangle+|1\rangle}{\sqrt{2}}$ from the state $\frac{|0\rangle-|1\rangle}{\sqrt{2}}$. The situation is easily remedied, if apply to the qubit A the Hadamard gate. You can check that

$$H|\Psi_+\rangle = \frac{1}{2}\begin{pmatrix} 1 & 1 \\ 1 & -1 \end{pmatrix}\begin{pmatrix} 1 \\ 0 \end{pmatrix} = \begin{pmatrix} 1 \\ 0 \end{pmatrix} = |0\rangle.$$

$$H|\Psi_-\rangle = \frac{1}{2}\begin{pmatrix} 1 & 1 \\ 1 & -1 \end{pmatrix}\begin{pmatrix} 1 \\ -1 \end{pmatrix} = \begin{pmatrix} 0 \\ 1 \end{pmatrix} = |1\rangle.$$

Therefore if after all these operations we measure the qubit A and find it in the state $|0\rangle$, the function $f(x)$ is constant, otherwise it is balanced. Note that we need only one measurement and that we never actually learn what is $f(0)$ or $f(1)$. We only learn whether they coincide or not.

Exercise

A qubit is in the state $|\Psi_+\rangle$. What is the probability of finding it in state $|0\rangle$? Or in state $|1\rangle$?

Then answer the same questions if the qubit is in the state $|\Psi_-\rangle$

The Deutsch algorithm is perfectly useless. Nevertheless it is a good and simple illustration of how quantum parallelism is used in digital quantum computing: the qubit input register is put in a superposition of all possible states, the computation is therefore performed on all these states at once, and then the result is cleverly manipulated in such a way that the contributions which correspond to the wrong answers (or to the information we do not need to know) cancel each other, and only the answer we want is encoded in the state of the output qubits and can be read out by measuring them.

The Shor algorithm, code-breaking and the promise of an exponential speed-up

The quantum computing research was greatly accelerated by the interest (as you have already guessed) from the military and intelligence community. The interest was aroused by the discovery in 1994 of the *Shor algorithm*[8], which would break the *public key cryptosystems*.

Key idea: Public key encryption simplified

A popular way of describing the public key encryption is to imagine that Alice maintains a stock of freely available, open strongboxes with a spring latch. Anybody can put a message in a box, close it and send to Alice, but only Alice has the key and can open the box.

The public key cryptosystems solve the problem, which we have described in Chapter 11 (how to safely exchange coded messages) in a very elegant way. Alice, who wants to securely receive coded messages from Bob without going through the trouble of using the one-time pads, generates a private key and a public key (two very large numbers). The public key is made public, and anyone (e.g. Bob) can use it in order to encrypt a message. But it can only be decrypted using the corresponding private key (which Alice keeps to herself). The clear advantage of such a system is that anybody can send messages to Alice any time, with the same degree of security. If Alice wants to respond to, e.g. Bob, she will use Bob's public key.

The public key approach is built on an exquisite piece of mathematics, which would take us far away from quantum mechanics, but its main idea is quite simple. There are mathematical operations, which are easy to do and hard to undo. For example, it is straightforward to multiply two prime

---

8 Peter Shor is an American mathematician who made important contributions to quantum computing besides the eponymous algorithm.

numbers[9], no matter how large. But it is hard to find the factors, given the product. This operation lies in the basis of public key cryptosystems (such as the popular PGP ('pretty good privacy'), those used by the banks and internet merchants to conduct money transactions, and of course the encryption systems used by large corporations and governments).

'Hard' means that in order to find the factors of an N-digit number, all known classical algorithms[10] would require the amount of time, which grows with N faster than any power of N (that is, e.g. as $\exp(N)$). In other words, they are 'exponential', and not 'polynomial', in N.

In practice, it may take decades or centuries in order to find the factors and break the code by running the existing algorithms on the existing classical supercomputers. What is even worse is that as soon as somebody gets sufficient computing power to break, say, a 128-digit key code, it is enough to switch to a 256-digit key in order to make your codes safe again (and such switching is quite easy and cheap). On the other hand, the very knowledge that somebody can break public key encryption in a polynomial time would have very deep consequences for the global economy, defence and politics.

If it turns out that somebody will be able to break this encryption in 50 years' time, we must start changing our procedures right now.
Attributed to an NSA employee attending one of the first post-Shor algorithm quantum computing conferences.

Therefore the discovery of the Shor algorithm, which enables the factorization of a large number in a polynomial time using a quantum computer, had a huge impact on the field. It hastened

---

9   Prime numbers, numbers which are only divisible by themselves and 1.

10   So far, there is no proof that there is no classical factoring algorithm, which would require a polynomial time. Finding it or proving its non-existence would be a great discovery.

the development of quantum communications, to have a totally unbreakable encryption scheme, just in case. (Now such schemes are in limited operation.). And, of course, it gave a great push to the development of quantum bits and quantum computers themselves.

The operation of the Shor algorithm is quite involved, but the idea is again to put the N-qubit input register of the quantum computer in the superposition of all possible states corresponding to the numbers from zero to $(2^N - 1)$,

$$\frac{|0\rangle|0\rangle...|0\rangle|0\rangle + |0\rangle|0\rangle...|0\rangle|1\rangle + |0\rangle|0\rangle...|1\rangle|0\rangle + |0\rangle|0\rangle...|1\rangle|1\rangle + |1\rangle|1\rangle...|1\rangle|1\rangle}{\sqrt{2^N}}$$

and then only keep the right one. Unlike the Deutsch algorithm, the Shor algorithm is probabilistic. This means that if we want to factorize some number M using this algorithm, the quantum computer will give a correct answer Q only with a finite probability P. But this is not really a problem. Factorizing is a so-called NP-problem (nondeterministic polynomial time), which essentially means that while it is hard to find an answer, it is easy to check whether the answer is correct or not. In our case we simply divide M by Q. If there is a remainder, we just rerun the algorithm and check again. The probability of *not* getting the right answer after n attempts is $(1 - P)^n$ and will very quickly go to zero as n grows.

Schrödinger's elephants, quantum error correction and DiVincenzo criteria

In order to run the Shor algorithm, a quantum computer must contain at the very least enough qubits to fit in the number to be factorized. These hundreds of qubits must be manipulated in a very precise way (we have seen that even a simple Deutsch algorithm requires a lot of manipulations with qubit states in order to extract the information), and all the time they must be maintained in a coherent superposition of many different quantum states. This is a notoriously difficult task. A superposition of two quantum states of a single or a few qubits, like $\frac{|00\rangle+|11\rangle}{\sqrt{2}}$, is routinely called a cat state, and as we know they are already very fragile. In a useful digital quantum computer we must deal with real 'Schrödinger's elephant' states and somehow ensure their survival during its operation.

The decoherence time of a quantum system generally drops fast as the size of the system increases, and the decoherence time even of a single qubit is not long enough to perform the necessary number of quantum logic gates. Therefore the quantum state of a quantum computer will be disrupted long before it will finish the assigned task, and the result will be garbage.

The designers of classical computers met with a similar problem. Even though classical bit states, zero or one, are much more robust than the qubit states, they are subject to noise and fluctuations. In order to prevent errors, each logic bit is encoded by several physical bits. This redundancy (like the 'Alpha, Bravo, Charlie', etc.) with the periodic error correction ensures the stability of information.

The simplest example of classical error correction is the three-bit majority rule. Each logic bit is encoded in three physical bits: $0 \to 000, 1 \to 111$. In the absence of noise all three bits must have the same value. From time to time the bits are read out and corrected. For example, $010 \to 000, 110 \to 111$. This scheme protects only against single bit-flips: if two bits were flipped between the check-ups, the majority rules will make the mistake. No error correction method protects against all possible errors (a complete physical destruction of a computer is also an error of sorts), but the ones employed in modern computers are more than adequate. Unfortunately these methods cannot work with quantum computers, because the intermediate states of qubits are quantum superpositions and will be destroyed by a measurement.

The goal of error correction is to restore the state of a computer to what it was before the error happened. As we know, the rigorous *no-cloning theorem* (Chapter 11) prohibits copying an unknown quantum state, and measuring a state almost always destroys it. It would seem that digital quantum computing is impossible.

Nevertheless quantum error correction is possible. You recall that using the Deutsch algorithm one can tell whether $f(0) = f(1)$ or not *without* getting any knowledge about the values of $f(0)$ and $f(1)$. If now a single logic qubit is encoded in several physical qubits (e.g. the three-qubit encoding, $|0\rangle_A \rightarrow |0\rangle_A |0\rangle_B |0\rangle_C, |1\rangle_A \rightarrow |1\rangle_A |1\rangle_B |1\rangle_C$), one can use a similar approach. Note that such an encoding means that a *superposition* state of a logic qubit, e.g. $\frac{|0\rangle_A + |1\rangle_A}{\sqrt{2}}$, corresponds to an *entangled state*, $\frac{|0\rangle_A |0\rangle_B |0\rangle_C + |1\rangle_A |1\rangle_B |1\rangle_C}{\sqrt{2}}$. The additional qubits necessary for the encoding (and for other operations of a quantum computer) – in this example, qubits B and C – are called the *ancilla* qubits.[11]

Suppose that the logic qubit is in some unknown, possibly entangled, quantum state, but one of the physical qubits may have flipped. Then there exists a specific – always the same! – sequence of quantum gates, which will factorize out the states of the ancilla qubits and make sure that neither of these qubits is in a superposition state. In the three-qubit case, this means that the quantum state of the system will be either $|\psi\rangle_A |0\rangle_B |0\rangle_C$, or $|\psi\rangle_A |0\rangle_B |1\rangle_C$, or $|\psi\rangle_A |1\rangle_B |0\rangle_C$, or finally, $|\psi\rangle_A |1\rangle_B |1\rangle_C$. Now the ancilla qubits can be measured without disrupting the state of the first qubit, and there can be only four outcomes of these measurements (called syndromes).[12]

In either case there is a known sequence of quantum gates, which would restore the logic qubit to its quantum state *before* the error. We did not violate any rule of quantum mechanics, since the state of the physical qubit A (and of the logic qubit ABC) remains unmeasured.

---

11   Latin for a 'maid'.

12   In case of the Deutsch algorithm we could call the state of the qubit A a syndrome: if it is 0, $f(0) = f(1)$.

Table 12.6 Three qubits, four outcomes

Syndrome:	This means that:
00	No error happened
01	Qubit B flipped
10	Qubit C flipped
11	Qubit A flipped

The three-qubit code does not protect against arbitrary errors. One needs at least nine-qubit code to be protected against all one-qubit errors, and the costs rise steeply (both in the number of ancillas and of additional operations), if you want to be able to recover errors, which involve more than one logic qubit. A rule-of-thumb estimate is that one needs from 1000 to 10,000 physical qubits per one logic qubit, and from 1000 to 10,000 quantum gates for quantum error correction per one quantum gate required for the quantum algorithm.

Therefore a digital quantum computer (or a *universal* quantum computer, as it is often called, because in principle it should be able to implement an arbitrary quantum algorithm) is a really expensive proposition, which must satisfy the *DiVincenzo criteria* to be operational.

Spotlight: DiVincenzo criteria

A universal quantum computer must have:

1 well-defined qubits;
2 qubits initialization in a desired state (e.g. in the state $|0\rangle|0\rangle|0\rangle|0\rangle...|0\rangle|0\rangle|0\rangle\rangle$);
3 a long decoherence time;
4 a universal set of quantum gates;
5 qubit-specific measurements.

The requirements are quite logical, but they interfere with each other and are generally hard to meet. For example, a universal set of quantum gates and qubit-specific measurements mean that control and readout circuits must reaching if not all, but many enough qubits, making the insulation of the system from the external world more difficult, endangering its fragile quantum state and reducing the decoherence time. The decoherence time must be long enough to perform the quantum error correction operations (that is, from 1000 to 10,000 quantum gates, depending on how optimistic you are), but the additional ancilla qubits drastically increase the size of the system and tend to shorten the decoherence time.[13] Therefore a *universal digital quantum computer* still remains a rather distant goal.[14]

Quantum slide rules: adiabatic quantum computing and quantum optimizers

While falling short of implementing the Shor algorithm and breaking the public key encryption, the developments in experimental quantum computing since 1999 were spectacular, especially in design and experimental implementation of different kinds of qubits and qubit arrays.

One way of going around the strict requirements of a digital quantum computer was to find a way to build something less demanding, but still useful. In a sense, a universal quantum computer 'à la DiVincenzo criteria' is a kind of quantum Pentium®. You cannot build a Pentium® with steam age technology, even if you have all the blueprints. But you can make a lot of slide rules and use them to develop your technology to the point, when a Pentium® becomes finally possible.

---

13   To give you an idea, at the present time, the number of quantum gates one can apply to a superconducting qubit before it decoheres is about a hundred and growing.

14   Assuming, of course, that there is no *fundamental* ban on large enough systems being in a quantum coherent state. It may be that the best way of finding it out is to try and build a quantum computer.

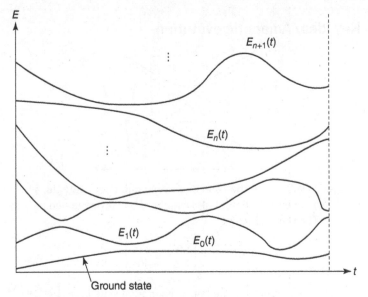

E

$E_{n+1}(t)$

⋮

$E_n(t)$

⋮

$E_1(t)$ $E_0(t)$

t

Ground state

Figure 12.5 Adiabatic evolution of energy levels

A similar approach in quantum computing is based on the *adiabatic theorem* of quantum mechanics. Suppose the Hamiltonian of a quantum system depends on time, so that its quantum state vector satisfies the equation $i\hbar \frac{d|\psi\rangle}{dt} = \hat{H}(t)|\psi\rangle$. At any moment of time the Hamiltonian will have a set *instantaneous eigenvectors*, such that $\hat{H}(t)|\psi_n(t)\rangle = E_n(t)|\psi_n(t)\rangle$. For example, its ground state will be also time dependent (Figure 12.5).

Then the theorem claims that *if* initially the system is in its ground state (i.e. $|\Psi(0)\rangle = |\psi_0(0)\rangle$), *if* the ground state of the system is *always* separated from the excited states by a finite energy gap (i.e. $E_1(t) - E_0(t) > 0$), and *if* the Hamiltonian changes not too fast (there is a rigorous expression for this 'not too fast', of course), then **the system will remain in its ground state** (i.e. $|\Psi(t)\rangle = |\psi_0(t)\rangle$) **at all times with the probability close to one.**

Key idea: Adiabatic evolution

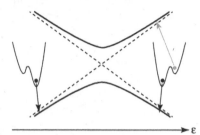

Adiabatic evolution in the case of a single qubit (for example, a particle in a two-well potential, Chapter 4) can be explained as follows. The Hamiltonian

$$\hat{H}(t) = -\frac{1}{2}\begin{pmatrix} \epsilon(t) & \Delta \\ \Delta & -\epsilon(t) \end{pmatrix}$$

depends on time through the bias ε. Its ground state is $\left|\psi_g(-\infty)\right\rangle = \begin{pmatrix} 0 \\ 1 \end{pmatrix}$ (i.e. in the right potential well) when $\epsilon = -\infty$ and $\left|\Psi_g(\infty)\right\rangle = \begin{pmatrix} 1 \\ 0 \end{pmatrix}$ (left potential well) if $\epsilon = \infty$.

If initially the qubit is in the ground state and then the bias changes slowly from minus to plus infinity, the qubit will have time to tunnel from the right to the left potential well, and will end up in the ground state. If the bias changes fast, the qubit will remain trapped in the right potential well, that is, in the excited state. (This 'trapping' is called the *Landau-Zener-Stückelberg effect* and is a major source of errors in adiabatic quantum computing).

Now, what does this have to do with quantum computing? The result of a quantum computation is a factorized quantum state of an array of qubits, e.g. $\left|1\right\rangle\left|0\right\rangle\left|1\right\rangle\left|0\right\rangle\left|1\right\rangle\left|0\right\rangle$, which can be read out one by one. It turns out that it is always possible to build an array of qubits, which has such a state as its ground state. It would seem then that all we need is to build such an array and cool it down[15] to almost absolute zero – and the answer will be ours without any trouble with quantum gates. Unfortunately, it is not that simple. Such an array will have a very complex *energy*

---

15 The scientific name for such cooling down is 'annealing'.

landscape – that is, there will be a huge number of states with almost as low energy as the ground state, and the system will almost certainly get stuck in one of these *metastable states*,[16] none of which is even close to our solution. This is where we use the adiabatic theorem. We will initially make the energy landscape of our system very simple – so that it will quickly cool down to its ground state, $|\psi_0(t)\rangle$ – and then will change the system very slowly. In the end we will be still in the ground state (with a probability close to one), but this ground state $|\psi_0(\tau)\rangle$ is now encoding the answer to our important problem. This approach is called the *adiabatic quantum computing*.

It has some clear advantages. Any quantum program can be run on an adiabatic quantum computer: the sequence of quantum gates can be replaced by a specific pattern of *static* qubit-qubit interactions. No need to apply quantum gates means that there is no need in a precision manipulation with quantum states of qubits. The ground state is the lowest energy state and if the conditions of the adiabatic theorem are satisfied, it is separated from the other states by a finite energy gap. Therefore as long as we maintain the temperature of our system smaller than this energy gap, suppress the noise coming from the outside and manipulate the system slowly (to avoid the Landau-Zener-Stückelberg effect), the system will not be disturbed, and its quantum coherence should be preserved.

Key idea: Quantum compass needles

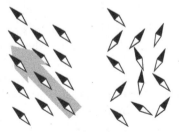

16　They are metastable, because eventually the system will go to the ground state – but it will take a practically infinite time.

The principle of adiabatic quantum computing can be illustrated as follows. Take a big set of 'quantum compass needles'. They will interact with each other and can make a number of metastable patterns. But in a very strong magnetic field they will all point in the same direction – this is our state $|\psi_0(0)\rangle$, to which the system readily goes. If now the magnetic field is very slowly lifted, the resulting pattern will be the one corresponding to the ground state $|\psi_0(\tau)\rangle$.

Not all these advantages are as clear-cut as one would like. The 'rewriting' of a circuit based quantum programme in terms of an adiabatic quantum computer requires some unrealistically complex patterns of couplings between qubits. There are reasons to believe that one cannot operate this kind of a device indefinitely long, since its quantum coherence will be eventually disrupted. The energy levels of such a device may not satisfy the conditions of the adiabatic theorem. But still, it is a tempting proposition, and there can be some short cuts.

One such short cut is a *'quantum optimizer'* or *'quantum annealer'*. This is a limited-purpose adiabatic quantum computer, a kind of 'quantum slide rule'. It is indeed like a quantum counterpart to a classical analogue computer. It would solve not an arbitrary problem, but a specific one, or a specific limited class of problems. It may not even solve them exactly, but give a good enough approximation. Many important problems either belong to this class (optimization problems) or can be reduced to them.

A good example is the *travelling salesman's problem:* given a set of towns connected by roads, find the shortest route passing through each town at least once. This is an important logistical problem, especially if you can solve it (or produce a good enough approximation) in real time. And this is an ideal problem for a quantum optimizer.

The machines produced by D-Wave Systems Inc. (Figure 12.6) belong to this class of quantum devices. They contain hundreds

Figure 12.6 D-Wave Systems quantum optimizer

(now even a couple of thousand) superconducting flux qubits, which are coupled to their neighbours in such a way that the coupling strength can be tuned. Their quantum states can be manipulated and read out using control circuitry, which runs from the outside to the inside of a dilution fridge (a special refrigerator, which can reduce temperature in a small chamber to less than 0.04 degrees above the absolute zero). Currently these machines are the biggest quantum devices in existence – so big, actually, that it is no longer possible to fully model their behaviour (as Feynman warned), and it is therefore very hard

Figure 12.7 The 128 qubit quantum processor produced by D-Wave Systems in 2010 shown against a slide rule to give a scale (left), and in a close-up (right). Photos by R. Plehhov

to tell, whether and to what degree they are really quantum devices. As of now the jury is still out – but the very fact that we can (actually *must*) discuss and test quantum devices on this scale is in itself extremely exciting.

Fact-check

1 The main difference between an analogue and a digital computer is that

   a a digital computer is built of electrical circuits

   b an analogue computer works faster

   c an analogue computer uses a physical process, which mimics the problem to be solved

   d a digital computer has an unlimited accuracy

2 The main disadvantages of analogue computers are

   a non-universality

   b limited accuracy

   c high maintenance costs

   d impossibility to program

3 A logic circuit is

   a a set of electronic circuits on a chip

   b a set of universal logic gates

   c a set of logical operations realizing an algorithm

   d a set of truth tables or Venn circles

4 Reversible gates are all logic gates, which

   a have the same number of inputs and outputs

   b can be represented by a truth table

   c can be represented by Venn circles

   d can be run in both directions

5 A set of universal quantum gates includes

   a all one-qubit and two-qubit gates

   b Pauli gates and CNOT

   c Pauli gates

   d Hadamard gate and CNOT

6 If the Hadamard gate is applied to a qubit in state $\begin{pmatrix} u \\ v \end{pmatrix}$, the result will be

a $\dfrac{1}{\sqrt{2}} \begin{pmatrix} u+v \\ u-v \end{pmatrix}$

b $\dfrac{1}{\sqrt{2}} \begin{pmatrix} u-v \\ u+v \end{pmatrix}$

c $\dfrac{1}{\sqrt{2}} \begin{pmatrix} u \\ -v \end{pmatrix}$

d $\dfrac{(u+v)|0\rangle + (u-v)|1\rangle}{\sqrt{2}}$

7 The Shor algorithm allows us to break

    a one-time pad encryption
    b public key encryption
    c Enigma code
    d Morse code

8 Quantum error correction is necessary in order to

    a break the no-cloning theorem
    b allow a quantum computer to operate longer than the decoherence time of a qubit
    c increase the accuracy of quantum computation
    d correct random errors during quantum computation

9 Adiabatic quantum computers

    a use the adiabatic theorem
    b are non-universal
    c are slow
    d are convenient for optimization problems

10 DiVincenzo criteria apply to

    a all quantum computers
    b digital quantum computers
    c analogue quantum computers
    d quantum optimizers

Dig deeper

M. Le Bellac, *A Short Introduction to Quantum Information and Quantum Computation*. Cambridge University Press, 2007.

A. Zagoskin, *Quantum Engineering: Theory and Design of Quantum Coherent Structures*. Cambridge University Press, 2011 (Chapter 6).

A. M. Zagoskin, E. Il'ichev, M. Grajcar, J. Betouras and F. Nori, 'How to test the "quantumness" of a quantum computer?' Front. Phys., 30 May 2014. Online at http://dx.doi.org/10.3389/fphy.2014.00033

13

History and philosophy

Quantum mechanics, together with theory of relativity, made a lasting impression on the way the public sees the operation and the meaning of science. It became the test study for the philosophers of science, who used the scientific revolution of early 20th century as a standard by which to evaluate not only the development of physics from Galileo's time to now, but the evolution of science in general. It is therefore worth our time now to have a brief look both at the general principles of science and at the events which shaped our current understanding of quantum mechanics.

In a certain sense, the 'Quantum Revolution' was an unfortunate choice of a standard by which to measure the development of human knowledge. As you have seen, quantum mechanics is far removed from everyday human experience and intuition, and can only be adequately expressed (and related to conventional, classical mechanics) in the specialized language of abstract mathematics. This makes it a very special case even inside the realm of science – and, as lawyers say, hard cases make bad law.

Quantum mechanics was also a very special case, because its discovery and early development seem having been propelled by a tidal wave of sheer luck. Looking at the history of science, it is hard to find another such example with the experimental evidence first literally forcing scientists to build from scratch an utterly counterintuitive theory, going against all previously held assumptions, and then confirming the theory whenever and wherever it was applied, and with an unbelievable precision. Special relativity comes distant second, in part, because – for all its 'paradoxes' – it is a very simple and straightforward theory; since Hermann Minkowski demonstrated that it is actually a bit of pseudo-Euclidean geometry, its consistency was never in any doubt and it never had to be changed in any significant way.

Sturm und Drang: the old quantum theory

To begin with, the Planck's formula of 1900 was almost literally dictated by the need to interpolate between Rayleigh–Jeans and Wien's laws for the equilibrium electromagnetic radiation. This in no way diminishes his great achievement and intellectual bravery. But it did certainly help that his formula, based on a simple assumption that electromagnetic energy is emitted or absorbed by matter in discrete portions, has as limiting cases the two formulas so well founded in classical physics. The Wien's law for the spectral density of *high*-frequency of electromagnetic radiation not only agreed with the experiment, but was derived from thermodynamical considerations. (Equilibrium thermodynamics is arguably *the* most reliable discipline in all of science.) The Rayleigh–Jeans law for the spectral density of *low*-frequency electromagnetic radiation was derived directly from Maxwell's well-established electromagnetic theory and statistical mechanics, and also agreed with the experiment. This clearly indicated that Planck's formula is not just an *ad hoc* hypothesis, and that it is likely to reflect some new aspect of reality.

Planck's formula did not say anything about the properties of light itself – it would be as correct if light was not quantized at all and only the matter had some non-classical properties. This got Einstein interested.

Even though beer is always sold in pint bottles, it does not follow that beer consists of indivisible pint portions.

Albert Einstein about the question of whether the light quanta exist, quoted in Philipp Frank, Einstein: His Life and Times (Da Capo Press, 2002), translated from the German by George Rosen.

In 1905 Einstein developed his quantum theory of the photoelectric effect. It was lucky that the effect was so well investigated by that time, and that its experimentally established laws were so simple and so impossible to reconcile with classical physics[1] – and, of course, that it turned out so easy to explain its properties assuming that light not only is absorbed and emitted in quanta, but also exists and propagates as quanta – in spite of its well-established wave properties! This was the crucial insight but also the beginning of the 'wave-particle duality', a redundant concept that still persists in many textbooks and encyclopaedias.

In 1909 Ernest Rutherford proposed his planetary model of the atom. It was based on incontrovertible experimental evidence and completely contradicted classical electromagnetic theory (because the orbiting electrons would have emitted radiation and dropped on the nucleus), mechanics, crystallography and chemistry (because, on the one hand, a classical electron can orbit the nucleus in any way it chooses, and so atoms could have any shape and size – and, on the other hand, chemistry and crystallography say that the atoms of the same kind should be all alike).

Then in 1913, Niels Bohr – who worked with Rutherford in 1911 – came up with *his* model of the hydrogen atom. Bohr *postulated* that there exist certain stationary orbits (*orbitals*), where electrons can circulate without emitting energy, and that electromagnetic energy is emitted and absorbed only when

---

1 Just imagine that instead of the photoelectric effect one had to start from superconductivity.

they jump from one orbit to another. The jumps were *totally unpredictable*, and the motion of an electron during the jump was *impossible to describe* by any trajectory in real space and time. The size of the orbital was determined by the demand that the angular momentum of the electron was quantized in the units of $\hbar$.

This was the foundation of the *old quantum theory*. This picture of fixed orbits and of indescribable and unpredictable quantum jumps, accompanied by the emission or absorption of light quanta, flew in the face of classical physics with its Laplacian determinism. However, not only did it resolve the problems with atomic stability and their fixed sizes, but it was in an excellent agreement with the spectroscopic data, expressed by such formulas as Balmer's and Rydberg's. So Bohr's theory was accepted, not as a finished theory, but as a working hypothesis, pointing at some unknown properties of Nature. This was lucky – for more complex atoms with more electrons, the Bohr's model is far from precise.

The old quantum theory was further developed by many physicists, notably by Arnold Sommerfeld (of the Bohr–Sommerfeld quantization). It was Sommerfeld – a brilliant mathematician – who worked out a more general model of hydrogen and hydrogen-like atoms, with not only circular, but also elliptic electron orbits not necessarily confined to the same plane, and thus discovered other quantum numbers – l,m – besides the principal quantum number n.[2]

The old quantum theory was not a consistent theory. It relied on the Bohr–Sommerfeld quantization rule, the adiabatic invariants and the *correspondence principle*, proposed by Bohr.

Spotlight: Correspondence principle

In the limit of large quantum numbers (or the Planck constant $h \to 0$) quantum formulas must reduce to classical formulas.

---

2   It is worth noting that the list of Sommerfeld's PhD and graduate students at the University of Munich reads as a 'Who's Who in 20th Century Physics', among them seven (!) Nobel Prize winners, including Heisenberg and Pauli.

This approach worked well enough with position, momentum and orbital angular momentum, but was not helpful at all if a quantum observable – like spin – did not have a classical counterpart. Old quantum theory provided solutions to a number of important problems, which turned out to be correct or almost correct, but its role was that of scaffolding. Unfortunately too much of this scaffolding survived in the public perception, popular literature and introductory textbooks, often in a very distorted shape, confusing and misleading the public – things like 'electron orbits' (which do not exist), 'quantum leaps' (which, besides being an obsolete concept, are exceedingly small – so a 'quantum leap' in anything happening on human scale means 'nothing happened'), and the standard picture of an atom, like the one on the cover of this book which has nothing to do with reality.

When all roads led to Copenhagen: creation of quantum mechanics

Quantum luck did not end there. In 1924 Louis de Broglie published his 'matter wave' hypothesis, extending the 'wave-particle duality' of light to all quantum systems. Even though the nature of a wave associated with a particle (e.g. an electron) was not clear at all, it gave a more intuitive explanation to the Bohr orbits (as those fitting an integer number of de Broglie wavelengths), and was promptly confirmed by the observation of electron diffraction off the crystal of nickel.

Not only was the hypothesis accepted, but it directly led to the discovery of the Schrödinger equation in 1925 (published in 1926): Schrödinger was going to give a talk at a seminar about this fresh new hypothesis. When in the process of working out what exactly would give rise to the de Broglie matter waves, he came up with his equation. Initially he, like de Broglie, worked within a relativistic theory and obtained an equation for the hydrogen atom, which was giving wrong answers. (The equation in itself was not wrong; it simply did not apply to electrons. It was later rediscovered and is now known as the *Klein–Gordon equation* for relativistic massive bosons.)

Fortunately, Schrödinger decided to try a non-relativistic approximation, which turned out excellent and produced the equation we know and use. This was the 'wave mechanics'.

At the same time – a few months in advance – Werner Heisenberg (and Paul Jordan and Max Born) worked out the 'matrix mechanics', which associated with every observable (as we call them now) a matrix. Essentially, it was a way of directly representing operators in Hilbert space, and the Heisenberg equations of motion were first written for these matrices.[3] The achievement is even more remarkable, since Heisenberg rediscovered the mathematics of matrices by himself. Obviously, the linear algebra was not a standard part of physics curriculum at that time. Matrices generally do not commute, and this directly led Heisenberg to the formulation of the *uncertainty principle*.

The same year Paul Dirac recognized the relation between the commutators and the Poisson brackets and demonstrated both the direct relation between quantum and classical mechanics and the equivalence of Heisenberg's and Schrödinger's approaches. Finally, in 1926 Max Born proposed the probabilistic interpretation of the wave function.

That was more or less it. The latter developments built on this structure; extending it toward the relativistic domain; applying to more and more complex systems; making it more elegant; looking for unexpected or unsuspected conclusions and applications; removing the scaffolding – but the structure of quantum mechanics was there to stay.

What the FAPP?

The questions about the *interpretation* of quantum mechanics arose from the very beginning. The very term 'interpretation' is revealing: there is no doubt that quantum mechanics gives the

---

3   You recall how Heisenberg and Jordan missed on David Hilbert's suggestion that their matrices may have something to do with differential equations!

right answers, but the intuitive understanding of why it happens is somehow hard to come by.

> *A scientific truth does not triumph by convincing its opponents and making them see the light, but rather because its opponents eventually die and a new generation grows up that is familiar with it.*
> Max Planck, *A Scientific Autobiography* (1949)

In the case of quantum mechanics, it was not as hopeless as Planck's saying seems to indicate. The new truth did convince the opponents and made them see the light. They just were rather uncomfortable with the truth – or maybe the light did hurt their vision – and this discomfort to a certain extent persists four generations later.

The two key features of quantum mechanics, its nonlocality and stochasticity, were to a certain extent the hardest bits to swallow. Einstein was very unhappy about them, but his attempts to demonstrate the incompleteness of quantum mechanics (in particular, the EPR Gedankenexperiment) eventually led to a direct experimental confirmation of these very features. They turned out to be the reflection of the fundamental properties of Nature, and not bizarre artefacts of quantum mechanical formalism.

The question of whether the state vector is 'real' begs the question of what 'real' means. Is it something you can touch or see? Then atoms or X-rays are not 'real'. Is it something that produces objectively observable and predictable effects? Then yes, state vectors are 'real'.

Can one say that rather than being 'real', a state vector just contains the information we can have about the physical system? Yes, but then it is necessary to say who are 'we' and what is 'information'. Information must have a material carrier, and information depends on the recipient. A letter in Klingon will not be all that informative to a hobbit.

There exist a number of interpretations of quantum mechanical formalism. I have mentioned earlier the 'Copenhagen interpretation', which rigidly separates quantum and classical worlds without telling you where to draw the boundary. You have certainly heard about the 'many-world interpretation' proposed by Hugh Everett; the 'pilot wave' interpretation by David Bohm goes back to Louis de Broglie himself. There are other interpretations. But the bottom line is that as long as one talks about 'interpretations', they all must lead to the *same* standard equations and rules of quantum mechanics and yield the same predictions. **FAPP – For All Practical Purposes – they are all the same.** I would therefore wholeheartedly support the dictum of physicist David Mermin: Shut Up and Calculate.

'But where is the physical sense'? Well, what *is* the physical sense, if not the intuition acquired by the experience of dealing with concrete problems based on a specific physical theory? In classical mechanics we deal with forces, velocities, accelerations, which are all vectors, without much fuss – but there are no 'vectors' in the Universe you could touch or point at. (The realization that in the material world there is no such thing as a 'triangle', a 'sphere' or a 'dodecahedron', goes back to the ancient Greeks, in particular Plato.) In quantum mechanics we deal with quantum states, which are vectors in Hilbert spaces, and observables, which are Hermitian operators in these spaces, and there is no reason to make fuss about that either. The main thing is that they correctly describe and predict the behaviour of the outside world, the *objective reality*, which is independent on anyone's subjective mind.

'But is this a *true* description?' Here we must talk about science in general, the scientific method and the philosophy of science.

'It was a warm summer evening in ancient Greece'

A single word 'science' means three interwoven and inseparable, but distinct, things: the system of *objective rational* knowledge about Nature, which has not just *explanatory*, but also *predictive* power; the system of *methods* for verifying, testing, correcting, systematizing and expanding this knowledge; and the human activity and men, women and societal institutions involved in this activity (who – as usual – spend too much of their time and energy on trying to achieve the optimal distribution of limited resources).

We will start from ancient history. It was the Greeks who realized that rationality and objectivity of any knowledge are closely interrelated. If you wish, this is due to our biological limitations. To begin with, Alice cannot see through the eyes of Bob nor have a direct access to his thinking process. Therefore Alice cannot directly check the veracity of Bob's statements – whether they concern solving an equation, observing a giant sea serpent or receiving a revealed truth – unless they are linked to some *evidence* accessible to Alice by an *explicit chain of reasoning* – that is, rational arguments. Moreover, subjective sensations may be deceiving, inner reasonings may be faulty. On the other hand, a rational argument based on evidence gives both Alice and Bob (and Eve, Festus, Goliath, etc.) a reliable instrument for testing their *own* knowledge about the world.

The central question here is, *what* rational arguments can establish the objective truth of a statement, given that we must rely on our imperfect senses, and quite often, faulty reasoning.

Plato, for example, believed in the inherent ability of a human mind to perceive the truth, and supported it by the theory of 'recollection'. A human soul by its nature belongs to the world of eternal, perfect, universal, immutable ideas (such as the idea of a perfect mathematical triangle), and therefore it has the intrinsic knowledge of these ideas, however distorted by the soul being incorporated in a human body. Therefore what we call acquiring knowledge is not getting something new. Instead it is recalling of what the soul 'knew but forgot'. From here it follows that the truth can be achieved by the process of pure reasoning as remote as possible from the world of rough and imperfect, accidental 'realities'. These 'realities' are, at any rate, only illusions, the deformed, fleeting shadows of true shapes on the cave's wall.

Key idea

The Allegory of the Cave from Plato's *The Republic* is arguably the most powerful and beautiful piece of philosophical writing in human history.

'Behold! human beings living in a underground den, which has a mouth open towards the light and reaching all along the den; here they have been from their childhood, and have their legs and necks chained so that they cannot move, and can only see before them, being prevented by the chains from turning round their heads. Above and behind them a fire is blazing at a distance, and between the fire and the prisoners there is a raised way; and you will see, if you look, a low wall built along the way, like the screen which marionette players have in front of them, over which they show the puppets.

I see.

And do you see, I said, men passing along the wall carrying all sorts of vessels, and statues and figures of animals made of wood and stone and various materials, which appear over the wall? Some of them are talking, others silent.

You have shown me a strange image, and they are strange prisoners.

Like ourselves, I replied; and they see only their own shadows, or the shadows of one another, which the fire throws on the opposite wall of the cave?

True, he said; how could they see anything but the shadows if they were never allowed to move their heads?

And of the objects which are being carried in like manner they would only see the shadows?

Yes, he said.'

Unfortunately, Plato's recollection theory fails to resolve the problems of, first, *how* an individual can distinguish between the truth or falsehood of his own reasoning; and, second, how to bridge the gap between individuals, if the criterion of truth is some 'inside feeling' of it.

This weakness was well understood by Plato's greatest disciple, Aristotle. To begin with, Aristotle turned around the Platonic argument about the relation of ideas (like a mathematical, perfect triangle, or an idea of a 'generic cat') and their realizations (like a triangle made of three pieces of wood or a particular Tabby). Instead of the latter being an imperfect shadow of the idea of a perfect cat, it is the general cat concept that is abstracted from the meowing particulars by the process of induction. Now Aristotle had to propose a different criterion of truth, and it was essentially the agreement with the experience: 'falsity and truth are not in things, but in thought'. Or, more specifically: 'It is not because we think truly that you are pale, that you *are* pale, but because you are pale we who say this have the truth.' This was a big step in the direction of the modern scientific method.

Aristotle did not deny the existence of absolute truths. To him (a rather unspecified) God was the source or embodiment of these truths as well as the Prime Mover for the physical world. But he proposed an explicit method of finding these truths by means of imperfect human senses and sometimes faulty human reasoning. While he could do little about improving the former, Aristotle revolutionized the latter by formulating the rules of logic. His revolutionary work in this area remained unsurpassed until the late 19th century. Its main significance for scientific method lies in the elimination of errors from the chain of reasoning, which binds any hypothetic statement to its directly observable consequences. Therefore, however long such a chain may grow, it still binds the consequence with the hypothesis. By disproving a consequence one thus overthrows the hypothesis – and whatever other statements are logically bound to it.

Due to Aristotle, science could become a much more rigid structure – more difficult to build, perhaps, but much more coherent and reliable. Of course, this development took quite a long time.

The proof of the pudding

The crucial importance of logical chains binding very remote statements for science did not become clear at once. For much of the time that passed since Aristotle those who had time, desire and opportunity to investigate Nature, more often than not based their investigation on some 'self-evident' truths, which could be directly extracted from either everyday experience or general ideas of what is 'true', 'proper', 'simple' or 'beautiful' (and more often than not, the outcome of the investigation should better agree with the locally preferred interpretation of this or that revealed truth). For example, the deeply rooted Greek belief that a circle is *the* perfect shape (being indeed the most symmetric and therefore the simplest to describe) forced astronomy into fifteen centuries of Ptolemeian contortions, only ending with Kepler's laws. The obvious facts that every inanimate object on the Earth will not move unless pushed, and will eventually stop, and that the celestial bodies, on the contrary, move eternally and regularly, produced the Aristotelian physics, which was only unseated by Galileo and finally demolished by Newton.

The rise of modern science was made possible by the recognition that one cannot demand 'simple simplicity' and 'obvious beauty'; that an observation or an experiment may need a painstaking interpretation, eliminating the side effects, noise, mistakes; and that the correct theory of an effect may be very far removed from everyday experience and common sense. In other words, modern science was made possible by the development of the *hypothetico-deductive method.*

Its essence is not far from Aristotle's approach. First we try to explain particular experiments or observations by creating a general hypothesis ('induction'). Then we derive from this hypothesis various consequences ('deduction') and test them. If the predictions fail, the hypothesis is rejected and must be replaced by a new one. If a hypothesis can explain several previously unrelated groups of experiments or observations and predict new phenomena (like the Maxwell's electromagnetic theory), then it replaces the particular hypotheses and becomes a **theory** (which in scientific usage means something as firmly established as it is at all possible in science).

The key difference is in the practical impossibility in most cases of *directly* testing a hypothesis and in the length and character of the chains of reasoning, which connect the observations to the hypothesis, and the hypothesis to its observable (and therefore testable) consequences. These chains more and more are forged from mathematics, and the hypotheses and theories are formulated in more and more abstract terms, further and further removed from our everyday experience. This is an objective process: science already reached far beyond our capacity for a direct observation. As Galileo pointed out, Nature speaks to us in the language of mathematics. It can be learned, and a long enough experience and training will make those of us who have the necessary 'linguistic abilities' quite fluent in it – but the process can be quite frustrating.

It was also Galileo who first fully recognized and forcefully expressed the futility of our attempts to make Nature fit our pet notions. The birth of modern science – or of science in the full meaning of this word – can be counted from the moment of this recognition.

> *...to prove that a thing is thus and thus, he resorts to the argument that in that manner it is conformable with our understanding, or that otherwise we should never be able to conceive of it, or that the criterium of philosophy would be overthrown. As if Nature had first made men's brains and then disposed all things in conformity to the capacity of their intellects. But I rather incline to think that Nature first made the things themselves, as she best liked, and afterwards framed the reason of men capable of conceiving (though not without great pains) some parts of her secrets.*
>
> Galileo Galilei, *Dialogue on the Great World Systems*

The full power of the hypothetico-deductive method was demonstrated in Newton's 'Principia': From observations (Kepler's laws) to the hypothesis (the Newtonian three laws of motion plus the universal gravity proportional to $1/R^2$) and then to testable predictions. Predictions were quantitatively confirmed, which validated the hypothesis and made it a theory.

This approach met a very serious opposition in continental Europe among the followers of Descartes. This great thinker built his philosophy on the concept of self-evident truths, and his physical theory around the notion of vortices in aether, a medium, which permeates the entire Universe and moves in circles, pushing bodies to each other. This intuitive, simple, easy to grasp picture had just one disadvantage – borrowing the phrase from Laplace, 'it could explain everything and predict nothing'. Newton's dismissive attitude to such 'hypotheses' was justified. Michal Heller formulated the crucial difference between the two approaches very well.[4]

---

4   A physicist in the field of general relativity and relativistic cosmology, philosopher of science and Roman Catholic priest.

> *The history of philosophy knew many various criteria of truth, and each of them, in the opinion of its adherents, should have been unfailing. Descartes was different from his predecessors only in modelling his criterion on the feeling of obviousness which is created by mathematical truths, and in subjecting it to a more rigorous, than others, critical analysis. But in the end his criterion shared the fate of all the earlier ones – it became a chapter in the history of thought. It is true, that Newton's criterion – agreement with experience – is 'sensual' and unambitious. It does not invoke any obviousness; on the contrary – sometimes in spite of the obvious it commands to accept what experience teaches. That is, it assumes the intellectual humility: the reality should not be suited to my intellectual capabilities (e.g., to what I consider obvious); it is I who must accept the reality the way it is and the way it opens to me in experience. An additional element of intellectual humility lies also in the fact, that I can see no necessary reason why such a strategy should work. It is simply so: from the moment when this method started being used, the history of science legitimates itself by an unparalleled row of successes.*
>
> Michał Heller, *Philosophy and Universe*

In the 20th century, Karl Popper – a philosopher of science – stressed that a scientific theory, in order to be scientific, must be *falsifiable*. This means that it must give such predictions, which can be tested. If a theory cannot be tested (either because it does not produce any predictions, or because its predictions can always be twisted to fit any observation or experiment), it is not scientific. In other words, a scientific theory can never be 'proven' (shown to be an eternal and immutable truth valid for all circumstances), but can only be *dis*proven, i.e. shown to contradict experiments and/or observations. On the other hand, no matter how many observations or experiments agree with a theory, there is no guarantee that next time it will not fail.

Of course, this principle was used in mathematics ever since the Greeks developed the concept of proof, as something quite obvious. A theorem is not proven by any number of examples, but can be proven false by a *single* counter-example. A finite number of examples – 'finite induction' – does not produce a proof. Examples, though, can direct us to a valid statement. This is what the reasoning by induction is about. Moreover, in mathematics, there is a formal procedure (mathematical induction), which allows statements to be proved based on what can be thought as an *infinite* number of examples.

The inherent 'intellectual humility' of the scientific method is sometimes taken as a sign of its weakness. Science does not produce eternal truths! Science is tentative! Science was wrong in the past, so it may go wrong again! In philosophy such views are often based on the so-called 'Neurath's boat argument':

> *We are like sailors who have to rebuild their ship on the open sea, without ever being able to dismantle it in dry-dock and reconstruct it from the best components.*
> Otto Neurath, 'Protocol Sentences', in A.J. Ayer (ed.) *Logical Positivism* (1959)

The boat is the corpus of scientific theories, and yes, it shows the signs of its origin, and is probably not the best one could build. In other words, the argument states that science cannot go beyond a set of makeshift 'models', and therefore does not tell us anything about the 'true nature of things'. But the argument actually defeats itself: however unsightly and imperfect, with all its wear and tear, this is still a boat. Its structure sufficiently reflects the reality of the outside ocean to keep afloat and move in the desired direction, and it allows further improvements. Even if one cannot get the 'ultimate' theory, we can tell which of the competing theories is better in describing observations and making predictions, and therefore can improve our knowledge. There is no logical proof that we will be always capable of getting closer and closer to the absolute truth, but is such a proof possible or even necessary?

There is also another important consideration, stemming from Bohr's correspondence principle in the old quantum theory. As you recall, it was the requirement that in the limit of large quantum numbers its formulas should reproduce the formulas of classical physics. But this principle can be extended to all of science:

Key idea: Correspondence principle

A more general theory must reduce to a less general *correct* theory in the area of latter's applicability.

This means, e.g. that quantum mechanics *must* reproduce the results of classical mechanics when describing classical systems; special relativity *must* reduce to classical mechanics for slow moving objects (compared to the speed of light); general relativity *must* yield Newtonian universal gravity for small enough and slow enough masses; electromagnetic theory *must* give correct results for the case of electrostatics; relativistic quantum field theory *must* boil down to non-relativistic quantum mechanics for a single slow particle, and so on. Generally, this is what happens and great effort is applied to the areas where the transition is not yet absolutely seamless (like in the case of quantum-classical transition, which we have discussed earlier). This gives science – as the corpus of knowledge and as the method of its improving and increasing – both stability and reliability.

We fully realize that our knowledge is relative and will never encompass 'all truth'; our models are provisional and will be replaced by something better. But all the same, our models are not just manipulations with symbols in order to fit extra epicycles to the Ptolemaic heavens, to be thrown away in a moment; our theories each time capture something of the essential properties of reality, and this something will survive in whatever new theories we develop. This is the way of science – maybe slow, maybe tortuous, but systematic and never giving back any terrain once conquered. It is not Alexander the Great's brilliant dash to the end of the world, to an early death and disintegration of the empire – but the steady step of Roman legions.

(To finish with another military analogy, the philosophy of science can be compared to a baggage train. It is certainly useful, and sometimes absolutely indispensable, but it should follow the army of science. If the baggage train gallops ahead of the troops, it may mean one thing only: the army is routed.)

At the end of the day, the validity of scientific theories is confirmed by their routine everyday applications. For example, a computer I use to write these words is the best proof that, e.g. quantum physics of solid state (microchips!) and light (lasers!) correctly grasps some very essential truths about Nature.

The final spell

On the one hand, science is now more powerful than ever, and its penetration in everyday life is ubiquitous. On the other hand, its frontier has moved so far away from common view, and its tools have become so sophisticated, that a lay person knows as much about modern science as the average Muggle knows about the curriculum of Hogwarts, though probably somewhat more than an Arthurian peasant knew about the activities of Merlin the Great.

This is not a comfortable situation. In Arthur's time peasants did not depend on Merlin's magic for their survival, and all the science and technology they needed was in their own heads and hands, or in those of the local blacksmith, carpenter and midwife. And the king certainly did not consult them about the size of Merlin's research budget. On the contrary, the very existence of modern society depends on the maintenance and development of its technology, which in its turn is not sustainable without living and active science. The decisions a society makes may be critical and they must be informed decisions.

Any sufficiently advanced technology is indistinguishable from magic.
Arthur C. Clarke, *Profiles of the Future* (1962)

This is why it is so important to understand why and how science works. It investigates Nature, and thus reflects some of Nature's ruthless beauty. Many things, which are considered important and *are* important when humans deal with each other, do not apply here. There is no freedom of opinion – once a scientific fact is established, you cannot choose to disbelieve it. There is no majority rule – one right opinion trumps no matter how many wrong ones. There is no consideration for feelings or mitigating circumstances – laws of nature cannot offend anybody, they are enforced equally and uniformly, and there is no Court of Appeal. The burden of proof is squarely on the person making a new claim. And the *only* thing that a scientist – as a scientist – cannot afford is to let his judgement be influenced by any consideration (however noble and altruistic) other than the search for a better understanding of the outside world.

Now let us return to quantum mechanics and the question of whether it tells us the truth. So far, all has been confirmed by the experiments and observations, so the question of whether the state vector is 'real' must be answered positively. It may be an uncomfortable thought, but I personally see nothing scary in the idea that what we see (including ourselves) are the shadows cast on the wall of a 'Hilbert cave' by a quantum state vector – very interesting shadows, which play with each other, influence the state vector and refuse to appear on more than one wall at a time. On the contrary, it is a fascinating picture. As Professor Dumbledore would have said, if something is happening in the Hilbert space, it does not mean that it is not real.

Dig deeper

H. Poincare, *The Foundations of Science: Science and Hypothesis, The Value of Science, Science and Method*. CreateSpace Independent Publishing Platform, 2014.

M. Kline, *Mathematics, The Loss of Certainty*. Oxford University Press, 1980.

M. Kline, *Mathematics and the Search for Knowledge*. Oxford University Press, 1985.

B. Russell, *Wisdom of the West*. Crescent Books, 1978.

Fact-check Answers

CHAPTER 1
1 (b)
2 (d)
3 (b) and (d)
4 (d)
5 (b)
6 (c) and (d)
7 (a) and (b)
8 (c) and (d)
9 (c)
10 (b)

CHAPTER 2
1 (a) and (d)
2 (a)
3 (a)
4 (b) and (d)
5 (c)
6 (b)
7 (a)
8 (b)
9 (b)
10 (a)

CHAPTER 3
1 (b)
2 (d)
3 (c)
4 (a)
5 (a)
6 (b) and (c)
7 (c)
8 (d)
9 (a), (b) and (c)
10 (b) and (c)

CHAPTER 4
1 (b)
2 (d)
3 (b)
4 (c) and (d)
5 (c)
6 (c)
7 (a) and (b)
8 (b)
9 (c)
10 (d)

CHAPTER 5
1 (a)
2 (a)
3 (b)
4 (b)
5 (b), (c) and (d)
6 (a) and (b)
7 (b) and (c)
8 (b) and (c)
9 (b)
10 (d)

CHAPTER 6
1 (a) and (b)
2 (c)
3 (a), (c) and (d)
4 (b)
5 (a)
6 (b)
7 (d)
8 (b)
9 (b)
10 (c)

CHAPTER 7

1 (a)
2 (b)
3 (b)
4 (c)
5 (b) and (d)
6 (d)
7 (a) and (c)
8 (c)
9 (d)
10 (a), (b) and (c)

CHAPTER 8

1 (b) and (c)
2 (d)
3 (c)
4 (d)
5 (d)
6 (a)
7 (a), (b) and (c)
8 (c) and (d)
9 (b) and (d)
10 (c)

CHAPTER 9

1 (c)
2 (a) and (b)
3 (b)
4 (a)
5 (c)
6 (c)
7 (b)
8 (b) and (d)
9 (b)
10 (d)

CHAPTER 10

1 (b)
2 (b)
3 (b)
4 (c)
5 (c)
6 (c) and (d)
7 (b), (c) and (d)
8 (b), (c) and (d)
9 (a) and (b)
10 (a)

CHAPTER 11

1 (b)
2 (b)
3 (b) and (d)
4 (a) and (b)
5 (a) and (d)
6 (c)
7 (a) and (b)
8 (d)
9 (a) and (d)
10 (b)

CHAPTER 12

1 (c)
2 (a) and (b)
3 (c)
4 (d)
5 (b)
6 (a) and (d)
7 (b)
8 (b) and (d)
9 (a) and (d)
10 (b)

Taking it further

Baggott, J., *The Quantum Story: A History in 40 Moments.* Oxford University Press, 2011.

Feynman, R. P. and A. R. Hibbs, *Quantum Mechanics and Path Integrals.* McGraw-Hill Companies, 1965 (there are later editions).

Feynman, R., R. Leighton and M. Sands, *Feynman Lectures on Physics*, Vols I, II and III. Basic Books, 2010 (there are earlier editions). See in particular Vol. I (Mechanics and some mathematics, Chapters 1–25), Vol. III (all of it – quantum mechanics) and Vol. II (Principle of least action – Chapter 19).

Gamow, G., *Thirty Years that Shook Physics: The Story of Quantum Theory.* Dover Publications, 1985.

Hamill, P., *A Student's Guide to Lagrangians and Hamiltonians.* Cambridge University Press, 2013.

Hawking, S. (ed.), *The Dreams That Stuff is Made of: The Most Astounding Papers of Quantum Physics – and How They Shook the Scientific World.* Running Press, 2011.

Kline, M., *Mathematics, The Loss of Certainty.* Oxford University Press, 1980.

Kline, M., *Mathematics and The Search for Knowledge.* Oxford University Press, 1985.

Lancaster, T. and S. J. Blundell, *Quantum Field Theory for the Gifted Amateur.* Oxford University Press, 2014.

Le Bellac, M., *A Short Introduction to Quantum Information and Quantum Computation.* Cambridge University Press, 2007.

Lipkin, H. J., *Quantum Mechanics: New Approaches to Selected Topics.* Dover Publications Inc., 2007.

Mattuck, R. D., *A Guide to Feynman Diagrams in the Many-body Problem.* Dover Publications Inc., 1992.

Peierls, R., *Surprises in Theoretical Physics*. Princeton University Press, 1979.

Peierls, R., *More Surprises in Theoretical Physics*. Princeton University Press, 1991.

Planck, M., *Scientific Autobiography*. Greenwood Press, 1968.

Poincare, H., *The Foundations of Science: Science and Hypothesis, The Value of Science, Science and Method*, CreateSpace Independent Publishing Platform, 2014.

Reif, F., *Statistical Physics* (Berkeley Physics Course, Vol. 5). McGraw-Hill Book Company, 1967.

Russell, B., *Wisdom of the West*. Crescent Books, 1978.

Wichmann, E. H., *Quantum Physics* (Berkeley Physics Course, Vol. 4). McGraw-Hill College, 1971.

Zeldovich, Ya. B. and I. M. Yaglom, *Higher Math for Beginning Physicists and Engineers*. Prentice Hall, 1988.

Ziman, J. M., *Elements of Advanced Quantum Theory*. Cambridge University Press, 1975.

Figure credits

Figure 4.16   Copyright © 2003, American Physical Society

Figure 5.6   © A. M. Zagoskin 2011, published by Cambridge University Press, reproduced with permission.

Figure 8.5   Quantum Theory of Many-Body Systems, Figure 2.1, 2014, p. 55, A. Zagoskin © Springer International Publishing. With permission of Springer Science+Business Media.

Figure 8.8   T.-S. Choy, J. Naset, J. Chen, S. Hershfield and C. Stanton. A database of fermi surface in virtual reality modeling language (vrml). *Bulletin of The American Physical Society*, 45(1): L36 42, 2000.

Figure 8.10   Quantum Theory of Many-Body Systems, Figure 2.3, 2014, p58, A. Zagoskin © Springer International Publishing. With permission of Springer Science+Business Media.

Figure 10.5   Quantum Theory of Many-Body Systems, Figure 4.5, 2014, p167, A. Zagoskin © Springer International Publishing. With permission of Springer Science+Business Media.

Figure 11.6   © A. M. Zagoskin 2011, published by Cambridge University Press, reproduced with permission.

Figure 12.2   © A. M. Zagoskin 2011, published by Cambridge University Press, reproduced with permission.

Figure 12.6   Copyright © 2011 D-Wave Systems Inc.

Index